THE WOMAN WHO CHANGED
HER BRAIN

Barbara Arrowsmith-Young is the Director of
Arrowsmith School and Arrowsmith Program.
She holds a B.A.Sc. in Child Studies from the
University of Guelph, and an M.A. in School
Psychology from the University of Toronto
(Ontario Institute for Studies in Education).

BARBARA ARROWSMITH-YOUNG

The Woman Who Changed Her Brain

How We Can Shape Our Minds and
Other Tales of Cognitive Transformation

WITH A FOREWORD BY
Norman Doidge, M.D.

VINTAGE BOOKS
London

Published by Vintage 2013

4 6 8 10 9 7 5 3

First published in Great Britain in 2012 by
Square Peg

Vintage
Random House, 20 Vauxhall Bridge Road,
London SW1V 2SA

www.vintage-books.co.uk

Addresses for companies within The Random House Group Limited
can be found at: www.randomhouse.co.uk/offices.htm

The Random House Group Limited Reg. No. 954009

A CIP catalogue record for this book
is available from the British Library

ISBN 9780099563587

Penguin Random House is committed to a sustainable future for
our business, our readers and our planet. This book is made from
Forest Stewardship Council® certified paper.

Printed and bound in Great Britain by Clays Ltd, St Ives plc

WITH GRATITUDE

Two people in addition to me were involved in the process of writing this book, and it would not be the book that it is without this collaboration. My heartfelt thanks to Annette Goodman and Lawrence Scanlan for each of the unique gifts you brought to the process.

Annette Goodman—for your collaboration in writing this book, for your gifted writing, and for your ideas that helped make it better than I had hoped, for helping to conceptualize the chapters at the outset, for identifying the key elements in each story to support the concepts being developed, for your quest through discussion and writing to find a way to make the ideas understandable and accessible, for your gift of finding the perfect flow for the ideas, for seeing how the pieces of the puzzle needed to fit together, for writing so beautifully about your own experience with learning disabilities, which richly contributed to illustrating those cognitive functions, and for your passionate commitment to alleviate human suffering and give children the tools to be whomever they choose to be in the world without the burden of learning disabilities.

Lawrence Scanlan—who rode the journey of this book, from the interviews of all the people who shared their stories, listening to and absorbing all of what they had to say—for finding the poignant beauty in the stories and committing them to paper, for your honed writer's craft and gift of finding just the right phrase or word to bring the material alive, for your ability to paint pictures with words that evoke the felt sense of the experience of having a learning disability, for unraveling the concepts in the science thereby making them intelligible, for showing me that sometimes less is more, and for your humor and patience throughout the process. Thank you for making the thoughts flow so eloquently onto the page.

NOTE TO READERS

For Aleksandr Romanovich Luria
(1902–1977)

VARIATION ON A THEME BY RILKE

(THE BOOK OF HOURS, BOOK I, POEM 1, STANZA 1)

A certain day became a presence to me;
there it was, confronting me—a sky, air, light:
a being. And before it started to descend
from the height of noon, it leaned over
and struck my shoulder as if with
the flat of a sword, granting me
honor and a task. The day's blow
rang out, metallic—or it was I, a bell awakened,
and what I heard was my whole self
saying and singing what it knew: I can.

—DENISE LEVERTOV

CONTENTS

Foreword by Norman Doidge, M.D.   xiii

Introduction   1

ONE   **THE ANATOMY OF RESISTANCE   9**

TWO   **MONAGHAN ROAD   15**

THREE   **LEARNING (AND REVERSING) MY ABCs   19**

FOUR   **THE FOG   23**

FIVE   **BRAIN WORK: ARROWSMITH CORE PRINCIPLES   29**

SIX   **UNIVERSITY HAZE   37**

SEVEN   **THE FOG IS DISPELLED   41**

EIGHT   **LOST IN TRANSLATION   45**

NINE   **HITTING THE WALL   57**

TEN   **WORDS FAIL   77**

ELEVEN   **LEAP BEFORE YOU LOOK   89**

TWELVE   **WHEN A PICTURE DOES NOT PAINT A THOUSAND WORDS   103**

THIRTEEN   **A CLOSED BOOK   117**

FOURTEEN   **NOTHING TO WRITE HOME ABOUT   131**

FIFTEEN   **BLIND TO ONE'S OWN BODY   145**

SIXTEEN   **A SCHOOL TAKES SHAPE   161**

SEVENTEEN   **LOST IN SPACE   165**

EIGHTEEN   **DRAWING A BLANK   177**

NINETEEN   **SEEING AND NOT SEEING   187**

TWENTY   **WHEN 2+2 DOES NOT EQUAL 4   191**

TWENTY-ONE   **IN ONE EAR AND OUT THE OTHER   201**

TWENTY-TWO   **THE IMPACT OF LEARNING DISABILITIES   207**

TWENTY-THREE   **WORD SPREADS   215**

APPENDIX 1   Description of the Cognitive Deficits Addressed
              by the Arrowsmith Program   217

APPENDIX 2   Lobes of the Brain   225

APPENDIX 3   Brodmann Areas of the Brain   227

              Notes   229

              Further Reading   237

              Acknowledgments   247

              Index   251

FOREWORD

For 400 years clinicians were taught that the brain was like a machine with parts. An electronic version of this metaphor is still with us when we think of the brain as a computer and are told it is "hard-wired," as though its circuits are finalized in childhood. Most clinicians trained in the second half of the twentieth century were taught a version of this model. Some still are.

This hardwired-machine model of the brain had devastating consequences for children and adults with learning disorders. It gave rise to a fatalism about their condition, which meant that they were *in all cases, necessarily,* condemned to live with their disabilities because machines can't rewire themselves. At best, we could teach these children to find ways to work around their problems.

About thirty years ago, a number of major neuroscience experiments were conducted that overthrew this view of the unchanging brain. Often they went unnoticed; sometimes, when noticed, disbelieving scientists trapped in the earlier machine model assumed that these experiments were based on sloppy methods, or that the results applied only to animals, or if to humans, only to small parts of the human brain. These experiments showed that the brain is neuroplastic, meaning that it is changeable, and that mental experience, and mental exercise, could alter its very structure.

It took twenty years for mainstream neuroscience to begin to accept that these experiments were sound and applied to humans, and not only to part of the brain, but to all of the brain, all of the time. Today we can say these experiments have been replicated thousands of times. Research and clinical trials throughout the world have shown that neuroplastic approaches can be used to treat traumatic brain injury, stroke, obsessive-compulsive disorder, learning disorders, pain, aspects of schizophrenia, and other afflictions. Neuroplasticity is suddenly much spoken of, is a "hot" term, and many marketers are putting old wine into new bottles—taking various simple brain games and rebranding them as "neuroplasticity exercises."

When tackling brain processing problems, however, as with so much else, the devil is in the details. One must have an intimate understanding of the pace at which the brain changes, how to "dose" the exercises, and which brain function to target. The latter is important because a simple problem, e.g., a reading problem, can actually be caused by a weakness in any number of different brain areas, and only one of these need be weak for a person to have a reading problem. So, what is required is not just an all-purpose brain exercise (which does not exist) but a brain-based assessment of the person's difficulties. These assessments and exercises often require years of refinement. Realizing that neuroplasticity has huge implications for education, neuroscientists at labs all over the world are getting their feet wet developing this work.

One woman began applying neuroplastic principles first to herself and then to students, just after the first experiments were done thirty years ago. The future in neuroplasticity arrived in a one-room schoolhouse in Toronto about a third of a century ago when Barbara Arrowsmith-Young and the team at her lab school began applying neuroplastic principles to learning problems. Barbara's own story—which I recounted in a chapter entitled "Building Herself a Better Brain" in my book *The Brain That Changes Itself* (2007)—and which is movingly elaborated in this book, is truly heroic, on par with the achievements of Helen Keller.

Barbara Arrowsmith-Young was born burdened with a number of extremely serious learning disabilities, including a severe inability to understand logic and cause and effect or to understand events in real time. When she read about lab experiments that demonstrated plasticity in animals given cognitive exercises, she began to develop her own brain exercises. This was astounding for two reasons. First, because she was able, despite her learning problems, to persist, reading difficult articles multiple times until she could break through her mental fog and understand them. Second, because she was able to use what she learned to create mental exercises that worked and lifted that mental fog once and for all. Usually in science, those who make breakthroughs in treating brain injury are fiercely intelligent people with extraordinary brains, working with those who have severely compromised brains. Arrowsmith-Young played both roles. And because she had been so disabled, she went on to develop numerous exercises for her other learning disabilities. At the end of this process, she found she was sufficiently equipped to open a school that could treat many of the major learning disorders.

Open since 1980, this school has now had more than thirty years to refine these exercises and to develop a brain-based diagnostic approach to learning disorders. To my knowledge, it is still the only school completely devoted to helping students, not to work around their brain problems— which is still standard practice in most schools—but to work through them, building up the students' relatively weak brain areas with brain exercises. It treats more learning disorders, to my knowledge, than any other school of its kind. As I envisage the future of neuroplastic education, I think that Arrowsmith-Young's notion of a school that has multiple brain exercises at its core for much of the day is the most promising model to get children back on track as quickly as possible.

This is not to say that everyone who has tried Arrowsmith-Young's exercises has succeeded; she has never made that claim, and would be wary of anyone so enthusiastic as to make it. No treatment works for everyone all the time, and I believe this is in part because our neuroplastic brains all develop differently, based on our genetics and experience. There always has to be some healthy tissue available for neuroplastic work to be done. In some brain-damaged children, the healthy tissue is very limited—but not in most. To this ideal school for learning disorders, I would add developments in neuroplasticity as they came along. Arrowsmith-Young's exercises are superb for dealing with cortical problems, but there are also new developments that address subcortical problems, and there is a role for neurofeedback and for sound-based interventions as well.

I can report having spent not several days but several years, many days a week, in the Arrowsmith School observing the results described in these pages, and I got to know many of the students described. Watching them over the course of several years, I saw them grow and develop. I observed their changing test scores, read the group data from the school, took the brain-based assessments, tried the exercises, and referred people, and I can attest to the remarkable progress that the substantial majority of students made. This is all the more impressive because most of them had previously tried, and failed, to progress using "compensations" that worked around their problems. I am not against compensations; often they work for people who have one or two areas of difficulty. But most children diagnosed with learning disorders have dysfunctions in a number of areas, as will be described in this book—even if they present with a single symptom, such as difficulty reading—and sometimes they don't have enough alternate brain areas to work around their problems. In these cases, they must use brain exercises to build up new processing areas. Another problem with compensations is that every time we choose to work around a brain area, we neglect it, further weakening what it can do.

I have used the terms *learning disability* and *learning disorder,* as does Arrowsmith-Young. I am mindful that in some areas there is a well-meaning movement afoot to end the use of these terms, an argument that they are destructive because stigmatizing. It is thought that renaming these children as "differently abled," or some such euphemism, will protect them. But I think the case histories in this book show that the suffering of these children is not in most cases caused by the stigma attached to their deficits (indeed, if anything, learning disorders still fly under the radar, are underappreciated, and are frequently misdiagnosed and medicated as "ADD"), but by the great difficulties they have processing, difficulties still poorly understood by many clinicians and educators alike. Indeed, I first realized how devastating these conditions are when, as an adult psychiatrist and psychoanalyst, I saw people who had had undetected learning disorders as children—and saw the developmental devastation those disorders had caused in terms of broken dreams, self-hate, depression, substance abuse. In statistical terms, the relationship of learning disorders to later mental health problems, substance abuse, job problems, and marital failure is frightening. Thus what is called for, to truly protect these children, is plainspokenness and proper help.

Indeed, much of what Arrowsmith-Young discovered about learning disorders or disabilities came from integrating the diagnostic concepts of the great Russian neuropsychologist Aleksandr Luria, who studied traumatic brain injury and brain damage. In fact, the effects of the most severe learning disorders are sometimes not so different from the damage caused by strokes and other kinds of brain injury (and some of the cases in this book are in fact of children who had brain injuries). I suspect that the attempt to rename these disabilities as merely "different" kinds of learning is born of despair, because many parents think there is nothing more that they can do for these poor children than "celebrate" their differences. True, many people do learn differently. But there is a difference between learning "differently" and being a child who must always struggle to learn, who is always falling further and further behind peers, no matter how much harder he or she tries. With programs like the one described in this book, we can focus on the right things and deliver better treatment to these children. Indeed, it wrenches my heart to know that we now have access to the kind of neuroplastic interventions that can help what I think is the majority of such children, but so few know about these methods so far. I can only thank my lucky stars that I live in the city where this school developed, and that I can refer the children I know, when appropriate, to Arrowsmith for help. It wrenches my heart equally to think of all the children, sitting in schools throughout the world, wiring into their brains each day the idea that they are dumb, or useless, or losers because many educators are still under the sway of the doctrine of the unchanging brain. I hope we don't have to wait for the usual generational change for this to be rectified.

Thus this book is an important document as much for educators as for children or adults with learning disorders and their family members. Reading it, along with Howard Eaton's *Brain School*, also based on the Arrowsmith Program, we now have a fuller picture of what goes on in such programs and a fuller picture of the kinds of conditions that can be helped. This gives us a model that can be imported into any school anywhere in the world, if the powers that be are willing to have their own special education teachers undergo some training. How much suffering would be relieved if only schools could begin doing the kind of brain-based assessments described here in the primary grades, to sort out which children might be helped.

I can't begin to describe the excitement I felt when I first met Arrowsmith-Young, this bold, ingenious, tormented, driven, deeply empathic pioneer. A whole new facet of human nature was revealed by her approach, and in many ways the scales fell from my eyes as I realized that one could understand one's own brain better by doing a kind of comprehensive brain-based cognitive assessment of oneself using Arrowsmith-Young's transformations of Luria's great discoveries. Even those without learning disorders could begin to understand the common "traffic jams in the brain" that are so common but that few have understood until now. Most everyone reading this book will find in its unique case histories a new way to think about people's difficulties in coping with the world. Here is an opportunity to understand the mental glitches and deeper problems of their own or of others in a new way.

Finally, this is a unique and very personal book. Arrowsmith-Young has been able to describe, in a poignant and often unforgettable way, what it feels like to have a devastating learning disorder—but also what it's like to leave it behind.

—Norman Doidge, M.D., author of *The Brain That Changes Itself*

The Woman
Who Changed
Her Brain

INTRODUCTION

March 2, 1943, Vyazma, Western Russia

On this sunny, almost warm but damp day, the soldiers are chilled, their army-issue felt boots soaked. Lieutenant Lyova Zazetsky, just twenty-three years old, commands a platoon of flame-throwers—part of a contingent pushing back against the German invaders who are dug in atop the steep and rocky banks of the frozen Vorya River.

Comrade Zazetsky looks west, where they will soon be headed. He talks to his men, encouraging them while they all wait impatiently in the stillness, as they have for the past two days. Finally, the order comes to advance, and the only sound he hears now is the clank and screech of armor stirring. In a low crouch, Zazetsky moves across the river ice at a pace between walking and running when the enemy begins to fire. As he hears machine-gun bullets whizzing over his head, he drops down instinctively under the hail of artillery. Then he rises and presses on. Then nothing.

Zazetsky's next memory is of coming to "in a tent blazing with light. . . . All I can remember is that the doctors and aides were holding me down. . . . I was screaming, gasping for breath. . . . Warm, sticky blood was running down my ears and neck. . . . My mouth and lips had a salty taste." A bullet has penetrated his helmet, then his skull, and has done massive damage to the left occipito-parietal region of his brain, leading to a prolonged coma and severely affecting his ability to reason. With damage to this area, the world of making connections and understanding relationships is lost. Even after hours of patient explanation, Zazetsky cannot fathom that an elephant is bigger than a fly (he knows that one is big and one small but cannot grasp the relationship between the two; the words *bigger* and *smaller* confound him).

Later he is shown photos of variously colored cats and asked to state which is bigger and which smaller. This too is beyond him.

"Since I was wounded," Zazetsky writes, "I've only been able to compare one word with another—one idea. And here there were so many different ideas that I got awfully confused." Unable to see the relationships between things, he sees the world as separate parts. Even something as simple as connecting the big and little hand on a clock is now impossible. He no longer understands logic, cause and effect, grammar, or dialogue in a film. For Zazetsky, the words in a movie come too quickly. "Before I've had a chance to figure out what the actors are saying," he writes, "a new scene begins."

Zazetsky, a gifted student with three years of study in a polytechnical institute behind him, takes months to grasp a basic element of geometry, only to have that hard-won knowledge vanish hours later.

The bullet had damaged the part of Zazetsky's brain that receives and processes input necessary for understanding the world. He could perceive properly with his eyes but could not deploy his brain to link perceptions or ideas, so he lived with disconnected elements. As Zazetsky put it in his diary, "I'm in a kind of fog all the time. . . . All that flashes through my mind are images, hazy visions that suddenly appear and disappear. . . . I simply can't understand what these mean."

He nevertheless writes a remarkable 3,000-page journal, gathered over the course of twenty-five painstaking years, in thick oilskin-covered notebooks. On some days, a sentence or two is all he can manage. "My memory's a blank," he writes. "I can't think of a single word. . . . Whatever I do remember is scattered, broken down into disconnected bits and pieces."

The damage to Zazetsky's brain is widespread and by no means confined to the area of the wound itself. His memory for information, for example, is severely damaged. Gone are the names of his mother and sisters and his address. He is unable to follow what he hears on the radio and gets lost on walks in the town where he was raised. Six years of studying German and three of English, advanced classes in chemistry: all utterly gone.

He holds a needle and thread in his hands and has a vague idea of their workings, but he can no longer summon the names of these and many other things. He urgently needs a bedpan, but he cannot summon that word. What comes to him instead are the words *duck* and *bird,* and he cannot decipher which is which.

Zazetsky has a handsome open face, with a strong nose and rugged black eyebrows, and at first glance he seems unscathed. But looks deceive. He can neither see nor imagine the right side of his body. Although he regains the ability to write (after six months of intensive schooling), the process is tortuous and slow, and he can neither read nor remember what he writes. He can speak, but only with great difficulty.

Worst of all, perhaps, is that Zazetsky is fully aware of his neurological deficits and is powerless to do anything other than to write about them in his own painful yet eloquent way.

"This strange illness I have," he writes, "is like living without a brain."

Late May 1943, Moscow

Zazetsky comes under the care of Aleksandr Romanovich Luria, a forty-one-year-old psychologist and a physician not long out of medical school. Luria heads a research team at a Russian army hospital looking at ways to help brain-damaged soldiers compensate for their neurological dysfunctions. In his new doctor, Zazetsky has two bits of good fortune. First, Luria's special and lifelong interest is aphasia—the difficulty speaking, reading, and writing that sometimes follows stroke or traumatic brain injury. Second, his brilliance is complemented by a rare compassion. Long after Zazetsky leaves the hospital, he and Luria remain close. They stay in touch for thirty years, meeting or speaking almost every week. A black-and-white photo of the two men shows them comfortably close together, each smiling at the other, Luria holding the fingers of Zazetsky's left hand ever so delicately in his own.

The writing of Zazetsky (a pseudonym) finds its way into a book that Luria writes in 1972, *The Man with a Shattered World: The History of a Brain Wound*. Zazetsky wants to call his writing *I'll Fight On*, and the title is a measure of the fierce resolve of this brain-damaged man to put the thoughts that come to him randomly into cohesive form. Zazetsky's writing is a desperate search for meaning, undertaken in the hope that his probing will help both himself and others—scientists studying the brain and those in circumstances like his own.

Each man helps the other. Had Zazetsky not crossed paths with Luria and been encouraged by him (the latter called his patient's writing "a triumph"), it's almost certain he would never have written his astonishing journal.

Luria is fascinated all his life by the brain (today he is considered a pioneer in neurology and the father of neuropsychology), and Zazetsky furthers his knowledge. Luria writes, "Precise knowledge was rarely to be found in the textbooks, which were filled with vague suppositions and fantastic conjectures that made maps of the brain scarcely more reliable than medieval geographers' maps of the world."

"His [Zazetsky's] description is exceptionally clear and detailed," writes Luria, and "if we follow him step by step, we may unravel some of the mysteries of the human brain." Through Zazetsky, Luria learned the geography and function of specific brain areas and made a major contribution to our understanding of the brain. The book you are now reading would never have been written had I not chanced across *The Man with a Shattered World* in 1977, the year Luria died. I shared Luria's intellectual curiosity and Zazetsky's reasoning deficit, as well as his determination. Zazetsky's drive led him to labor all that time writing a journal as he strove to understand the "strange illness" that had suddenly and catastrophically befallen him, leaving him with a loss of meaning in his world. My own drive compelled me to search for a solution to the same neurological deficit that had robbed me of meaning since birth.

Our shared determination, I would later understand, was actually a shared strength in frontal lobe functioning, that part of the brain critical for planning and seeking solutions. A hallmark of good functioning in this region of the brain is driven determination in pursuit of a goal.

Peterborough, Ontario, 1957

Six years old, I hear an exchange that fills me with a quiet horror. I have accompanied my mother to an after-school parent-teacher meeting to discuss the teacher's concerns about my slow progress.

"Barbara," the teacher is explaining to my mother, "has a mental block." As children do, I understood this truth quite literally. Evidently there was a chunk of wood lodged in my brain, and it would have to be removed.

The teacher was almost right. The word *block* missed the mark, but *blockage* was pretty close. For the first twenty-six years of my life, and I am fifty-nine years old as I write this, I lived in a dense fog not unlike Zazetsky's.

I too could make no sense of the relationship between the big and little hands of an analogue clock. Asked to perform the simple addition of a two-digit column of numbers, I would randomly choose numbers from the left or right side. The logic of basic math, the concept of telling time, the ability to truly comprehend what I was hearing or reading: all eluded me. On the playground, I couldn't follow conversations or the rules of simple games.

Depending on which question was asked on a test, I might get a grade of 29 or 92. What allowed me to progress through primary school, high school, university, and even graduate school were some exceptional strengths. My auditory and visual memory ranked in the 99th percentile (as a teenager I could watch the TV news at 6:00 P.M., and at 11:00 P.M., I'd parrot the broadcast as if I had the script in front of me). I also possessed exceptional mental initiative to attack and solve the problems that came my way, which translated into a singular work ethic and gritty determination to succeed.

My teachers' opinions of me varied widely. I was labeled "gifted," "slow," and "difficult." Some parts of my brain responded like a finely tuned musical instrument; others could not be relied on. There was no language then to describe my condition. The phrase *learning disabled* was coined only in 1962, by a Chicago psychologist named Samuel Kirk, and it did not come into common parlance until the late 1970s. Fifty years ago, when I was a child, students were seen as smart or slow or somewhere in between.

The educational system of the 1950s appeared to make up its mind about me early on. In the primary grades in those days, students were grouped with others who read at the same pace. I was put not with the "squirrels" (the quick readers), where I longed to be, and not the "rabbits" (the average readers) either, but with the "turtles" (the slow readers), who were mocked and teased by the other children. To my dismay, my reading problems were a result of letter and word reversals, which I could do nothing about. Almost universally assumed at the time was the idea that you had to play the hand you were dealt because the brain you were born with was fixed and hardwired. Period. A certain prevailing fatalism meant that I was told I had best learn to adjust.

My woes did not end there. As with Zazetsky, other areas of my brain were compromised. I took forever to learn how to tie my shoelaces, I was always getting lost, and I could not tell my left hand from my right. I constantly ran into things and bruised my body, chipped my teeth, and had stitches because my whole left side felt alien to me. I was "accident prone," but there was a reason for that and my other woes, and it had everything to do with my brain.

Photographs of me at the time show a handsome child, long-haired and freckled, as you might expect of someone with my mixed Scottish, Irish, and English heritage (my forebears had come to North America in the early 1600s). But my smile then was always closemouthed, and there was something quite muted about me, tentative and shy.

Teachers and even my own friends and family had no real sense of the anguish my learning challenges caused me and how hard I had to work to maintain my grades. And as I advanced from grade to grade, the going got harder and I had to double and redouble my efforts.

Ahead would lie periods of despair. By my teens, suicide seemed to me the only option.

Toronto, Ontario, 1977

When I was twenty-five years old and in graduate school, I happened upon Luria's *The Man with a Shattered World* and began reading Zazetsky's account of his life. As I read his words—"I'm in a kind of fog all the time. . . . All that flashes through my mind are images, hazy visions that suddenly appear and disappear"—I was dumbstruck. This brain-damaged soldier was describing himself, but he was also describing me. *I am Zazetsky,* I thought. *Zazetsky is me.*

The giveaway was the story about the clocks. Trauma inflicted on a particular part of someone's brain appeared to result in that person losing the ability to tell time. If Zazetsky was the man who couldn't tell time in postwar Russia, I was his female counterpart in Canada a few decades on. But where a bullet had inflicted the damage on this soldier's brain, I entered the world with my brain already damaged. Our problems had dramatically different origins, but their outcome was precisely the same.

I finally had an explanation for what had ailed me all my life. Here was evidence that my particular learning disabilities were physical, with each one rooted in a specific part of my brain. This realization marked the turning point in my life.

By reading Luria's books, *The Man with a Shattered World* and *Basic Problems of Neurolinguistics,* I came to understand that for both Zazetsky and me, the primary problem lay in the left hemisphere at the intersection of three brain regions: the temporal (linked to sound and spoken language), the occipital (linked to sight), and the parietal (linked to kinesthetic sensations). This is the part of the brain necessary for connecting and relating information coming in both from the outside world and from other parts of the brain in order to process and understand it. Both Zazetsky and I saw perfectly well and heard perfectly well; making sense of what we saw and heard was the issue.

As long as I live, I will never forget the palpable excitement I felt as I read Luria for the first time. Every page of his books offered revelations that I underlined and reread.

"The bullet that penetrated this patient's brain," Luria wrote, "disrupted the functions of precisely those parts of the cortex that control the analysis, synthesis, and organization of complex associations into a coherent framework."

Zazetsky and I could not make meaningful connections between symbolic elements, such as ideas, mathematical concepts, or even simple words. As he put it, "I knew what the words 'mother' and 'daughter' meant but not the expression 'mother's daughter.' The expressions 'mother's daughter' and 'daughter's mother' sounded just the same to me." I too, could not grasp the difference between "father's brother" and "brother's father" even when such language could be mapped onto concrete experience (my father did indeed have a brother).

Both Zazetsky and I caught fragments of conversations, but we never grasped the whole. The words came too quickly for us to decipher their meaning. My habit had been to replay—as many as several dozen times—simple conversations, the lyrics of a song, the dialogue in a movie as I strove to understand. But how could I understand even one sentence? I was still working on the meaning of the first part of the sentence and missed what came after. Logic, cause and effect, and grammar befuddled me, just as they had Zazetsky.

During this time, I came across the research that an American psychologist, Mark Rosenzweig, at the University of California at Berkeley had conducted with rats. He demonstrated that the brain can physically change in response to stimulation. *If a rat can change his brain,* I thought, *perhaps a human can do the same.* I married the work of Rosenzweig and Luria in order to create an exercise to change my brain.

The exercise, I knew, would have to be central to the function of my brain's particular weak spot. If my brain, for example, had trouble interpreting relationships, would rigorous practice interpreting relationships over a sustained period of time address the problem?

I had no idea whether this might work, but I had nothing to lose but time. And this I had already lost. Luria explained that people with lesions in this cortical region (the juncture in the brain of the parietal-occipital-temporal lobes) had difficulty telling time on an analogue clock. I wondered if a clock-reading exercise might stimulate this part of my brain.

I created flash cards, not so different from the ones my mother had used with me in first grade to teach me number facts. But this wasn't rote. This was me in 1978 at the age of twenty-six trying to activate a part of my brain that had never worked properly. Since I could not accurately tell time, I had to use a watch and turn the hands to the correct time (with a friend's help), and then draw the clock face. I would do the exercise every day for up to twelve hours a day, and as I got better at the task, I made the flash cards more complex, adding more, and more challenging, measures of time. They were relational components.

I threw myself into the exercise, as is my style. My brother Donald used to call me "an engine without a regulator."

The name of the game was speed and accuracy. How quickly could I calculate time—first simple time, then complex time? By gradually speeding up the exercise and making it harder, could I go from not being able to tell time to being better at it than the average person? If this worked—if I could get faster and more accurate at processing relationships on the clocks—then I had some hope that the related symptoms clustered in this impaired part of my brain might likewise improve: my inability to comprehend written material, my woeful grasp of math, my general lack of understanding in real time.

I cannot describe my exhilaration when I began to feel the result of all this work. Points of logic became clear to me, and elements of grammar now made sense, as did math. Conversations that I had always had to replay in order to comprehend now unfolded for me in real time. The fog dissipated and then lifted. It was gone for good.

What had happened? The part of my brain that was supposed to make sense of the relationship between symbols—most famously in my case, the hands of a clock—had been barely functioning. The work I did with flash cards activated that moribund part of my brain, getting the neurons to fire in order to forge new neural pathways. This part of my brain had been asleep for the first twenty-six years of my life, and the clock exercise had woken it up.

And what about my other issues: my klutziness, my penchant for getting lost? Did these problems have their origins in my brain, and could they too be helped or even eliminated by stimulating different parts of my brain? But which parts? And what exercises? This became my quest: to use what I'd learned from this experiment to wake up other areas of my brain.

What I have learned by doing this work for some thirty-four years is this: just as our brains shape us, we can shape our brains.

CHAPTER ONE

THE ANATOMY OF RESISTANCE

Why are educators still telling parents that learning disabilities are lifelong? Given the great weight of evidence for neuroplasticity, why are cognitive exercises not more widely recognized as a treatment for learning disabilities?

We now take it as a given that the brain is inherently plastic, capable of change and constantly changing. The human brain can remap itself, grow new neural connections, and even grow new neurons over the course of a lifetime.

When I went to university in the 1970s, I was taught that the brain was fixed: what you were born with is what you lived with all your life. This belief that a learning problem is a lifelong disability had major implications for education and learning. Education was about pouring content into a fixed system—the brain. At one point, it was argued that there were critical periods in childhood when the brain could more efficiently learn; once this window closed, such learning became more difficult. At best, then, the brain was seen as a fixed system with brief periods of malleability.

I remember attending a lecture in the late 1980s and being told that children with learning disabilities could be likened to different animals with various strengths. The eagle could soar and see the world from on high, the squirrel could run fast and climb trees, and the duck could gracefully swim in the lake. We were then admonished: never make the duck try to climb or the eagle to swim or the squirrel to fly. Find each child's unique gifts, we were told, and work on developing them because children could deploy them to compensate for things they could not do.

My own education had been grounded in this approach. And I knew from my own experience that the enormous expenditure of energy made in attempting to work around problems generated limited results.

Norman Doidge, the author of *The Brain That Changes Itself*, argues that centuries of viewing the brain as a machine, rather than an organ capable of regenerating itself, gave rise to what he calls neurological fatalism: the belief that to be born with a learning disorder was to live with it until death.

Presuppositions in any field (mine happens to be school psychology) determine how we carry out our investigations and what we believe is possible. Those presuppositions shape our view of reality and can become entrenched as truth, rarely to be questioned.

This, too, is neuroplasticity at work: we all create a map of how the brain works—a map based on our knowledge and training. Many people have not yet formed or understood the new map of the neuroplastic brain, especially in relation to education.

Doidge describes what he calls "the plastic paradox." The property of plasticity can give rise to both flexible and rigid behaviors. Because trained neurons fire faster and clearer signals than untrained neurons, when we learn something and repeat it, we form circuits that tend to outcompete other circuits. Soon there is a tendency to follow the path most traveled. If your occupation is offering remedial programs, this means: "We've always done it this way; let's continue doing it this way." Once a way of thinking and practicing within a framework becomes habitual, it becomes ingrained, and a significant amount of energy is required to reshape old thought patterns and institute new practices.

Although we now know that age, training, and experience make for a constantly changing brain, many educators have yet to learn how to deploy the principles underlying neuroplasticity (that is, to treat learning disabilities). Even educators who recognize that the brain is changeable are still engaged in professional practices based on the old brain-is-fixed paradigm. Certainly it takes time, effort, and learning to integrate new knowledge into common practice; meanwhile, most treatments for children with learning disabilities remain based on those old notions of hardwired brains and lifelong disabilities.

Thomas Kuhn, in his classic work published fifty years ago, *The Structure of Scientific Revolutions*, explains how the process of discovery works in science and what happens when there is a paradigm shift. Every field of science has foundational beliefs that people within that field learn as part of what Kuhn calls "educational initiation that prepares and licenses the student for professional practice." These beliefs and assumptions determine what is to be studied and researched within that scientific discipline. Research within the paradigm is designed to gather knowledge within the framework of the paradigm. In the process of research, as Kuhn describes it, anomalies emerge that cannot be explained by the paradigm's assumptions. At first, these anomalies are ignored or resisted. Over time, it's recognized that they violate the paradigm and need to be investigated. Finally, the old paradigm begins to shift, and the one that emerges encompasses the anomalies. Kuhn argued that a paradigm change is in essence a scientific revolution, and that the new scientific theory demands rejection of the older one. In this way, science develops. Neuroplasticity is one such new paradigm.

What we urgently need now is a new paradigm in education—one that crosses the great divide between neuroscience and education. This new model will wholeheartedly embrace the life-altering concept of the changeable brain and use the principles of neuroplasticity. The end result will be a fundamental change in the learner's capacity to learn.

Harvard University has developed the Mind, Brain, and Education Institute, devoted to bridging the gap between neuroscience and education. Its goal is to connect the disciplines; bring together educators and researchers to explore the latest research in cognitive science, neuroscience, and education; and apply this knowledge to educational practice. To help advance this goal, the institute also publishes a journal, *Mind, Brain, and Education.*

In an article published in fall 2010, "Linking Mind, Brain and Education to Clinical Practice: A Proposal for Transdisciplinary Collaboration," authors Katie Ronstadt and Paul Yellin note: "Increasingly, neuroscientists are identifying the neural processes associated with brain development, the acquisition of academic skills, and disorders of learning. Integrating this emerging knowledge into education has been difficult because it requires collaboration across disciplines." Part of the challenge, they note, is that neuroscientists and educators have different languages, frameworks, and priorities.

I started Arrowsmith School in Toronto in 1980. It evolved from my experience using the principles of neuroplasticity to address my own learning problems. I had become increasingly aware that traditional methods of dealing with learning-disabled students had only limited success. The Arrowsmith Program was developed from research in neuroscience, not education. The fundamental premise of my work is that by changing the brain, the learner's capacity to learn can be modified.

The principle of neuroplasticity is considered part of the field of neuroscience and has not traditionally been taught in teachers' colleges or studied widely in the educational system. Teachers who become administrators are taught that their job is to teach content. Thinking about rewiring the brain (so that the student becomes more capable of learning content) marks a radical departure from their traditional job description.

When I started this work more than thirty years ago, neuroplasticity was being discussed and researched in laboratories, but it was neither widely known nor well accepted. Only since 1990, partly encouraged by President George H.W. Bush's proclaiming the 1990s the Decade of the Brain, has neuroplasticity been investigated extensively. I vividly remember standing on Yonge Street in Toronto outside my school in May 1999 as I excitedly told a colleague about an article I had just read: "New Nerve Cells for the Adult Brain," by Gerd Kempermann and Fred H. Gage in *Scientific American*. This marked the first time I became aware of not just neuroplasticity but neurogenesis—how the adult brain can actually grow new neurons in the hippocampus, an area of the brain important for memory and learning. The brain was more plastic, more malleable, than originally thought.

Only as recently as 2000 did Eric Kandel of Columbia University win the Nobel Prize for his work demonstrating that learning in response to environmental demands changes the brain. Here was more proof of neuroplasticity. After Kandel won the Nobel Prize, it took several more years for the concept to reach the mainstream through media attention. Only in the past few years has the idea become broadly accepted in theory. In terms of the history of science and the acceptance of ideas, this is a fleeting moment.

Santiago Ramón y Cajal (1852–1934), considered one of the great pioneers in neuroscience, theorized the concept of neuroplasticity long before we had the refined technology and techniques to demonstrate it. He hypothesized, but could not prove, that the brain can be remapped, its very structure and organization changed, by the right stimulation. "Consider the possibility," he once said, "that any man could, if he were so inclined, be the sculptor of his own brain, and that even the least gifted may, like the poorest land that has been well cultivated and fertilized, produce an abundant harvest." This Spanish neuroscientist and histologist (one who studies the microscopic structure of tissue) won the Nobel Prize in 1906. Almost a century later, Kandel's work confirmed Cajal's hypothesis that the brain is plastic and changes occur at the synaptic connections between neurons.

The terms *neuroplasticity* and *brain plasticity* might feel new, but that's because it is only recently that these terms have gained currency. In fact, these terms have been around a long time, and research in neuroplasticity—though mostly on the margins, it must be said—has been under way for more than two hundred years.

In 1783, an Italian anatomist named Michele Vincenzo Malacarne studied the impact of exercise on the brain. He took pairs of birds from the same nest and subjected one pair to intense training, the other pair to none. He then conducted the same experiments with dogs: one pair got the enrichment of intense training, and the other pair got no stimulation. When the animals were euthanized, Malacarne found that the brains of stimulated animals were larger than those of their counterparts, and especially in the cerebellum—the part of the brain that governs motor control and coordination. And 165 years later, Jerzy Konorski, a Polish neurophysiologist, used the terms *brain plasticity* and *neural plasticity* in a book he wrote in 1948: *Conditioned Reflexes and Neuron Organization*.

Today neuroplasticity is generating a lot of excitement in areas of rehabilitative medicine, where good news is rare. Norman Doidge chronicles in one of his documentaries some of the promising research being conducted. Jeffrey Schwartz, an associate professor at the UCLA School of Medicine in California, for example, is using what he calls "self-directed neuroplasticity" in treating obsessive-compulsive disorder (OCD). The classic example of OCD is the person who can neither stop thinking about germs nor stop washing his hands to kill germs. Schwartz is deploying the principles of neuroplasticity to forge new pathways in his patients' brains. His patients are learning firsthand that the brain can change its structure in such a way that the impulses can be recognized as just that—mere impulses. The physiological changes that accompany this mental shift are visible on their brain scans.

Alain Brunet, an associate professor in the Department of Psychiatry at McGill University in Montreal, is using the malleability of the human brain to treat people suffering from posttraumatic stress disorder. These are victims, for example, of rape, child abuse, car accidents, and hostage takings for whom the event remains very much alive in their minds. Brunet is reporting success using a blend of pharmacology and neuroplasticity. These patients are first given medication to dampen the emotion associated with these memories and then asked to repeatedly recall the event. These men and women are rewiring their brains, disconnecting the circuitry linking the memory of the event to the arousal of their own threat systems. This process allows each person to file the memory in a new folder in the brain, not in the virtual present but in its rightful place—in the actual past. This is the principle of neuroplasticity in action: neurons that fire apart, wire apart. These new treatments for trauma usefully exploit this fact: when you remember a traumatic event, the network for that memory enters a more malleable state, and the treatment proceeds in that heightened neuroplastic milieu.

Finally, researchers in California are using cognitive exercises to help those with schizophrenia address some of the cognitive problems associated with their condition. Such people have difficulty perceiving, processing, and remembering information, and neuroscientists Sophia Vinogradov and Michael Merzenich are using specially designed computer programs to improve these cognitive functions. Brain imaging, their research shows, has demonstrated that these cognitive exercises change regions of the prefrontal cortex—those involved in regulating attention and problem solving—of a person with schizophrenia so it begins to look more like a normal brain.

In addition, a protein in the brain called BDNF (brain-derived neurotropic factor, also known as the "brain's fertilizer") is typically low in the brains of those with schizophrenia. Critical for neuronal survival, BDNF is also believed to play a vital role in what neurologists call activity-dependent plasticity (a term used to describe the brain's ability to change as the result of specific sustained stimulation). These exercises increase BDNF levels to normal—further evidence of neuroplastic change.

"We know the brain is like a muscle," says Vinogradov. "If you train it in the right way, you can increase its capacity. The brain is ever changing in relation to what's happening to it. With the correct training, we can improve cognitive processes that weren't strong to begin with by improving the processing pathways." Says her colleague, Dr. Merzenich, "The brain changes—physically, chemically, functionally."

"It's unrealistic," Norman Doidge told me recently, "to expect that the definitive demonstrations of neuroplasticity in the laboratory will suddenly undo the doctrine of the unchanging brain that so many were taught. Intellectual revolutions require time to spread. In the meantime, those few who have understood that neuroplasticity has immediate applications face incredulity or even opposition. That is what happens when you are at the cutting edge. It's lonely out there. But a lot of the opposition to the idea will pass generationally because in the last few years, all the major neuroscience texts have chapters on neuroplasticity. I'm not worried about its clinical acceptance in the long term."

CHAPTER TWO

MONAGHAN ROAD

1951

I came into the world on Wednesday, November 28, at 9:15 A.M. at Women's College Hospital in Toronto, Ontario. My mother was fine, but her baby daughter was asymmetrical.

"The obstetrician must have yanked you out by your right leg," my mother would later joke. My right leg was longer than my left, causing my pelvis to shift. My right arm was at an odd angle and never did straighten, and my right eye was more alert than the left. My spine was twisted, leaving me with a mild degree of scoliosis. My tiny body was flawed from birth, as was my brain, though that wasn't immediately apparent.

Two brothers, Donald and Will, followed me, joining our older brothers, Alex and Greg. Over the course of ten years, my mother, Mary Young, bore five children. She was a teacher, a nutritionist, and a school trustee. She and my father were founding members of the Unitarian Fellowship, and she helped design the curriculum for the church's Sunday school. My mother was a woman bent on saving the world—committee by committee, project by project. She was progressive in her thinking and, as a school trustee, backed innovations such as enrichment programs and French as a second language. My mother was a doer and a leader.

We would say the Unitarian prayer of grace before every meal, holding hands around the table and thanking God for the food we were about to eat but also invoking the notion of social justice and service to others. One of my father's favorite expressions was this: "What have you done today to make the world a better place?" Implicit in this was that each of us was put on earth for a purpose, and that purpose was to do some good.

My father, Jack Young, was an inventor by nature, a thinker, and a problem solver. He worked as an engineer at the General Electric plant in Peterborough, where we had moved from Toronto before I attended primary school. He was soft-spoken and quiet, brilliant and work obsessed, and my habit as a child was to wave good-bye to him as he headed off to work. Even when he was home, he often worked at a street-facing desk, near the fireplace, in the living room. His challenge at work was to convert electricity from one form to another to power huge electric motors. At one point, his task was to enable trains' smooth and efficient use of electricity. To this day, when I board a train and it accelerates smoothly, I think of my father and silently thank him.

Ironically, for all my brothers' keen desire to build their own machines and gadgetry, I was the one most interested in my father's inventions, and he would show them to me. I rarely grasped what they were all about, but I caught his passion and excitement for the creative process.

My paternal grandmother, Louie May Arrowsmith, was born in Provo, Utah, but in 1891, when she was eight, the family embarked on a year-long trek north by covered wagon to Creston, a town in the British Columbia interior they helped found and where my father was born. My middle name was originally Macdonald (in homage to the family connection with Canada's first prime minister), but when my brother Donald came along, he was given this as his first name and my middle name became Arrowsmith. On reflection, I found that this new name sat well with me, for I feel a deep and abiding connection to the spirit of my pioneering grandmother.

Among my first memories is myself at three accepting a dare to jump over the Christmas tree my father had tossed into the backyard. I fell short and landed in the middle of the tree, leaving my father to pick pine needles out of my face. I look back on that incident and I completely understand what that little girl could not have done: she could never have cleared the tree. Owing to a severe neurological deficit, the entire left side of my body was like foreign territory. It was as if I had suffered a stroke at birth.

Another early memory: I am three and playing in the driveway. As children do, I had invented a game—matador and bull. I was the bull, the car was the matador, and the game was to charge my parents' car in the driveway at full tilt and then swerve at the last moment. Instead, I ran headlong into the car and needed several stitches. I had misjudged how fast I was going and my body's position relative to the car. I remember sitting in that same car holding a towel to staunch the bleeding as my mother drove me to the hospital. It was the first such trip but by no means the last.

My mother turned to me as she started the car. "I'll be surprised," she said, "if you live another year."

Eventually I would come to understand the reason for my klutziness. I had a combined neurological deficit—one kinesthetic and one spatial— and until they were addressed, the accidents and mishaps would continue.

CHAPTER THREE

LEARNING (AND REVERSING) MY ABCs

1958

The grade 1 report card I took home in June 1958 points to my early academic struggles: "Barbara lacks self-confidence. She is very hesitant about answering and reading." There were no checks under "Excellent" or "Very Good." I'm not sure how I managed to earn "Good" in arithmetic, not when the comment in the margin was this: "Unsure of numbers, e.g., 3, 2, etc." My printing was only "Fair": "Does not stay on the lines properly."

I'd put up my hand to go to the bathroom and spend forty-five minutes in there to escape reading or arithmetic. Before that year was out, my mother had instituted a lunchtime regimen that involved the two of us meeting at home for twenty minutes of flash cards:

$$2 + 2 = ?$$
$$4 + 1 = ?$$
$$6 + 3 = ?$$

My mother printed the question and answer on opposite sides so she could keep track. I persuaded her to sit in the brightest part of the house during these tutoring sessions, and for some reason I seemed to be smarter in the sun. My mother eventually figured out that I was using the bright light to read the answers through the card, so she put her thumb over the answer.

I was doing what all learning-disabled children do: finding clever ways around my disability. My mother's campaign poster from 1978 listed her many accomplishments as school trustee under the subheading "Capable, Committed and In Touch." The poster noted that she had served seven years as a school volunteer under the auspices of psychological services— "to help children with learning difficulties." The one child with learning difficulties she had not been able to help was her own daughter.

My notebooks, which my mother kept, show mirrored letters and numbers and addition exercises where the teacher has crossed out many of the results with a large X. Wrong, wrong, wrong. That year, I had a meltdown in class. I sobbed inconsolably and pounded my head on my desk. The teacher had a responsible student take everyone else outside (where they gathered at the first-floor window and watched the drama unfold). My mother was called to the school.

This particular teacher was new to her profession and interpreted my confusing 6's and 9's and b's and d's as disobedience. I once got the strap in front of the entire class, and as painful as this was, the humiliation was worse. The teacher's other recourse was to have me write over and over again without reversing letters. I couldn't do that, and the failure made me feel helpless. Because I often wrote from right to left, my sweaty palm, moist from all the anxiety I was feeling, smeared the page. This only enraged the teacher further. Again, she read this as willful disobedience.

There I was in first grade, and I had already become a workaholic. That's what it took to get me through my first year in school.

Queen Mary Public School was across the street from our house. My brothers could have tossed a baseball and hit its red brick wall.

As I moved through the grades, school became more and more the source of my anguish. I continued to reverse letters and numbers. (But now and again, reversing numbers worked in my favor. Asked to add 12 + 13, I would add 21 + 31, get an answer of 52, reverse that number, and get a rare correct answer: 25.)

Some days I just couldn't face another class. There's a reference on my third-grade report card to "Barbara's illness," but my real affliction was pretty simple: I loathed school. So I would feign illness and put the thermometer next to a lightbulb to convince my mother that I was running a temperature.

I was struggling in English and natural science, and my report card reflected that. In arithmetic, the teacher's comment in the margin was this:

"Trying hard to overcome counting"—as if counting were a mountain to climb, and it was for me.

By fourth grade, the teacher's comments were getting testier: "Arithmetic problem-solving is extremely weak. She is very slow at all written work—most careless and untidy too. Composition needs careful attention." Attendance showed a whopping thirty-nine days absent from school.

In fifth grade, math and reading comprehension were still difficult for me, and the teacher suggested in her remarks that drills and repetition—adding numbers, memorizing the multiplication tables—would set everything right. She was wrong about that, and the pattern continued through elementary school.

By eighth grade, I had learned the correct math procedures but still needed extra time. In that class we were doing "rapid math," which provoked terrible anxiety in me and which I would do almost anything to avoid. The task in this exercise was to complete an entire page of calculations in five minutes. I would sneak the book home the night before, answer all the questions, and then write the answers in my book lightly in pencil; in class the next day, all I had to do was copy over my answers. This was the only way I could complete the exercise in the allotted time. Though I had done the work (at home), I felt like a fraud.

My brother Donald has a memory, a painful one, of me trying to learn. Dad's in the living room attempting to explain simple mathematical concepts to his only daughter, and he's almost in tears because his otherwise bright child cannot grasp what he is saying. I wanted to please him, I desperately wanted to understand, and sometimes I pretended that I did understand. Donald says he had to leave: it was too wrenching to watch father and daughter failing to connect across that great divide.

CHAPTER FOUR

THE FOG

1965

My father adored me, his only daughter, and without meaning to, he put pressure on me.

"I have only one daughter and four sons," he told me when I was thirteen, "and you've got to really make it." He was paying me a compliment, but I took him to be laying down expectations I could not hope to meet. Clearly his favorite child (my brothers concede that point), I really did not want to let him and the rest of the family down, so I determined to work that much harder to succeed. But given all the challenges I faced, I did wonder how I could manage. My struggle to learn became more fervent, driven by a fear of disappointing everyone.

By the time I entered high school, I was becoming ever more aware that I didn't comprehend ideas as others did. I was increasingly being asked to reason and think logically, but I could do neither. Symbolism, metaphor, historical cause and effect, mathematical and chemical equations: grasping these ideas was like trying to grasp a beam of light.

Understanding, really understanding, a newspaper article or a current affairs documentary on television remained beyond me. I would read an article and have no idea what the author intended, so I would reread it five or ten times. But I was never completely certain. Unable in class to discern what anything meant or what was important to document and what was not, I simply wrote down everything the teacher said.

I would easily feel overwhelmed when trying to follow a group discussion on any topic more complex than the weather. Slow to grasp meaning, I was always five steps behind everyone else. My outstanding memory allowed me to replay the dialogue later and only then be in a position to offer my two cents' worth—several hours too late.

I had even greater trouble grasping ideas when they came from more than one person. There was the general thrust of the conversation, then someone else's insight, then another's. It was all too much. The new ideas to be integrated were like foreign invaders. I had to hold on for dear life to what I knew, all of which might be threatened by these alternative ideas. Dizzy and disoriented by the effort of trying to follow and hold it all together, I knew that if I loosened my grasp to integrate this new information, everything would fragment and I would be lost. I was forever juggling balls in the air, but I could juggle only one at a time. Small wonder I was seen as inflexible.

The relationship between words stymied me, but so did individual words. Asked in a math class to multiply or divide, I struggled to know what I was being asked to do. I would use different colored markers on those operational words as guides—red, say, for dividing and green for multiplying. The colors were more useful to me than the words themselves. For the same reason, I would have to hunt around in my brain for the right words after saying, "I need to find the vacuum cleaner to cut the grass." The missing words, of course, were *lawn* and *mower.*

Later I would learn the term for this: *semantic paraphasia* (*paraphasia* meaning an error in naming). A person with this disorder substitutes one word for another word similar in meaning. I did this often. I would ask one of my brothers to help me fix my *radio* when I meant *tape recorder.* I knew the difference between these two objects, of course, but the wrong word flew out of my mouth. The words were tied only loosely to their referents, and one word would crowd out another.

One important signifier of the *symbol relations deficit* (as I would come to call this neurological weakness) is that you can't paraphrase what you've just heard or read. To put something into your own words requires that you comprehend someone else's words first.

In literature class, I could memorize from my notes that the great white whale in Herman Melville's novel *Moby Dick* symbolized an elusive, unattainable goal, an obsession that reached destructive proportions. But to me, the whale was just a whale.

And the link between the words *mammal* (a class of animals distinguished, in part, by the fact that females suckle their young) and *mammary glands* (the organs that produce the milk) never registered despite the common root. My world comprised unrelated facts—to be memorized but never understood.

Living in dense fog was one image I used to describe my confusion. Another image I used was cotton candy; I felt as if I was encased in sticky spun sugar that obscured my ability to see the world clearly, touch it, engage with it, fathom it. Until my twenty-seventh year, there were no *aha* moments when the various parts of something I had read or heard coalesced. I got pieces, context, and a general sense of things but never a logical conclusion—not one I could rely on. In the same way, I couldn't see logical inconsistencies in what others were saying, and that left me vulnerable to con artists. Nor could I understand jokes, so I learned to laugh when others did.

Double negatives eluded me. Sentences such as, "I am not unfamiliar with his work" or, "I do not disagree" made no sense to me no matter how long I pondered them.

"The boy chases the kangaroo" was, to my ears, pretty much the same as "The kangaroo chases the boy." I would have to create a picture—actually on a page or in my head—to verify meaning. My notebooks were filled with drawings, a graphic illustration of my ongoing attempt to understand language.

When someone spoke to me, I relied on his or her facial expression and intonation to get a general sense of what was being said, but I was never sure.

In class, how could I possibly know the answer when the question itself was invariably unclear? On exams, I was never sure I had interpreted the question correctly, and when I wrote my answer, I could never be sure I had conveyed the meaning I intended.

I would wait anxiously for my results, knowing that my grades could range from failure to 90 percent. What saved me from outright failure was my powerful memory, which allowed me to regurgitate course material without understanding it. My grades were invariably all over the map, and my teachers could only conclude that if I had excelled once, I could do it again. In their eyes, there was only one possible conclusion: I had not studied hard enough.

Studying for exams was a grueling experience—like trying to swim through quicksand. I had my rituals. One was to go into the basement and pound my head against the dryer. Maybe I was trying literally to pound some sense into that head of mine. Pulling hair out was another response. Or I would spread out my books on my bed and weep from the depths of my soul. Finally, I would seek out Star, the family cat. "Mr. Cat," as my brothers called him, would listen patiently as I poured out my misery.

Of course, I didn't know this at the time, but my own body was imposing its strategy, draining the amygdala of fear and emotion so I could concentrate on my work. (The amygdala, two almond-shaped nuclei located deep in the brain's temporal lobes, plays a role in storing and processing emotions. And nothing stirred up more negative emotion in me than studying for exams. Sheer blind terror was what I felt as I set out to accomplish a task I knew was impossible.) Mine was an instinctive outpouring of grief, maybe in hopes that with some of the anguish dispelled, I might better grasp all that I was supposed to. I never did. I could remember, and memory got me through school and into university. But I could not understand.

I felt alienated and out of touch with friends and classmates. My fragmented view of the world led to a fragmented sense of self. I developed a very negative self-concept and low self-esteem. I became depressed. I was enveloped in a fog that never cleared.

By the age of fourteen, I was so distraught that I harbored thoughts of suicide. I wanted an end to the emotional pain and exhaustion, the constant confusion and struggle. I took a razor blade and lightly cut both wrists, thinking I would go to sleep and not wake up in the morning. Next morning I berated myself for not getting even this right.

By this time, I had become increasingly slow to trust. I had few friends; in fact, I could cope with only one person at a time. Social encounters that most people would look forward to held terror for me. I knew I would not understand the conversations and would only be able to sit quietly hoping no one would try to engage me, to elicit a comment. A party was hell. Like Zazetsky, I could discern that I did not fit in, but I could do nothing to change this reality. The few friends I did have, I relied on if, say, I had to buy a calculator. A purchase meant choosing, weighing options. I found all of that extremely hard, but just as hard was relying on friends to perform this service for me. I didn't like having to depend on others.

I had written the following entry in my diary, trying to make sense of my experience: "When someone offers me advice, I can't be certain if it is appropriate for me or not. I just can't figure it out. I become rigid and cling to my decision; it has meaning for me and I have worked hard to understand it. I cannot let go of that security. When someone asks me to do something, I have trouble not just figuring out the 'why' but also often the 'what.' My world is so confusing."

I was different, but I had no idea why. My world consisted of a series of seemingly random events, ones I could not understand and over which I had no control. I was clinging to a cliff by my fingernails, and the only questions were these: When would I fall? Who would catch me? The answers, apparently, were "Soon" and "No one."

My mother described me in the Young family Christmas newsletter in 1966 as "methodical and conscientious." As best I could, I was covering my tracks.

CHAPTER FIVE

BRAIN WORK: ARROWSMITH CORE PRINCIPLES

In 1977, when I began exploring neuroplasticity, it was terra incognita—certainly in education. Now it is undisputed that the brain is plastic, malleable, capable of change. This is the biggest discovery about the brain in the past four hundred years, lifting us out of what Norman Doidge calls "the dark ages of neuroplasticity" and overturning centuries of conventional wisdom that the structure of the brain could not be modified.

In 2000, Eric Kandel won the Nobel Prize for demonstrating synaptic plasticity—that is, the strengthening of the connections between neurons as a result of exposure to stimulation, which resulted in learning and the formation of long-term memory. This was Hebb's rule (a concept introduced in 1949 by Canadian psychologist Donald Hebb) in action. Simply put, the rule is this: neurons that fire together wire together. The more often they fire together, the stronger their connection.

Mark Rosenzweig, whose work with rats so inspired me, demonstrated neuroplasticity in the late 1950s and early 1960s. In his scientific paper "Effects of Environmental Complexity and Training on Brain Chemistry and Anatomy," he summarized his findings, which clearly showed that rats housed in cages with toys, ladders, running wheels, and tunnels—what he called an "enriched" environment—performed better on maze tests than those in sterile cages, or "impoverished" environments. On autopsy the enriched rats were found to have heavier brains.

Subsequent animal studies replicated Rosenzweig's findings. The stimulated brains of these rats demonstrated a wide range of changes: more glial cells, which play an important role in neurotransmission at the level of the synapse by helping to establish and maintain synapses; larger capillaries, enabling greater blood flow to the brain; and an increase in enzymes involved in the synthesis and breakdown of neurotransmitters. The brains of stimulated rats also show an increase in gray matter (which contains both glial cells and dendrites) as well as denser synaptic connections through increased branching on dendrites and more spines on dendrites—the branches of the nerve cells that receive impulses. Simply put, the stimulated rat brain is a better brain with a greater capacity to learn and problem-solve. These neurological changes occurred regardless of age: the same results were found if the exposure to the stimulating environment occurred in infancy, childhood, or adulthood.

What I found most exciting in Rosenzweig's work was this: differential stimulation led to differential effect. In other words, if a rat was blindfolded and put in a tactile environment, the part of the brain related to touch was what changed. This finding became central to my work.

Stimulation of the brain led to physical and chemical changes that translated into improvements in learning. And if the stimulation was modified to place more demands on one region of the cortex than another, change would occur in the targeted part of the brain.

Such physiological changes are more difficult to measure in humans because noninvasive techniques such as brain scans must be used. But in several studies, scientists have been able to measure an increase in gray matter as a result of specific and intense forms of learning. For example, cab drivers in London, who must learn a massive number of routes before being licensed, have significantly more gray matter than is typical in the right hippocampus, an area related to spatial navigation (more on this in Chapter 17). And individuals who practice meditation show increased gray matter in an area of the brain linked to emotional regulation. Jugglers show increased gray matter in areas of the brain related to visual and motor activity.

To understand why an increase in gray matter might be important, let's look at a simplified version of how neurons communicate in the brain. Between each neuron is a gap called a synapse, and at the center of the neuron is an axon. Reaching out from each neuron are dendrites at one end and an axon terminal at the other (see Figure 1). An electrical impulse fires down the axon, causing a chemical called a neurotransmitter to be released from the axon terminal into the synapse; that same chemical is then received by the dendrites of adjacent neurons. These signals can be inhibitory or excitatory. If the net sum of all of the signals exceeds zero, the process continues with the signal being sent on to other neurons.

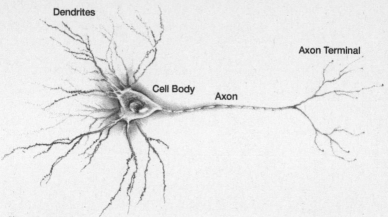

Dendrites

Axon Terminal

Cell Body   **Axon**

Figure 1. Neuron.

So why are more glial cells and dendrites good for neurotransmission?

Glial cells are involved in the communication process in the brain. They surround neurons, providing nutritional support by anchoring the neuron to its blood supply; they regulate chemicals in the synapse by removing excess ones and recycling others; they control blood flow to regions of brain activity; and they communicate with each other and with neurons.

Dendrites provide surface area on the nerve cell to receive signals coming from other neurons. More dendrite branching means more potential connections, allowing more signals to be received. In addition, recent research indicates that dendrites also release neurotransmitters.

Figure 2 (next page) depicts two neurons—one from an animal raised in an environment with stimulation and the other from one raised without stimulation. The more stimulating the environment, the more dendrites are created, allowing for more communication between neurons within the brain. This all adds up to better brain function.

Rosenzweig had a sense forty-five years ago of the extraordinary potential of neuroplasticity. "I hope," he wrote in 1966, "that the research we have been considering may eventually add, not cubits to stature, but cubic centimeters to human brains." When I read those words, I felt that a gauntlet had been thrown down. I resolved to create exercises to help individuals with brain deficits, or learning disabilities. I was determined to address the weak cognitive capacities that underlie learning dysfunctions. I wanted to address the root cause of learning disabilities rather than manage their symptoms.

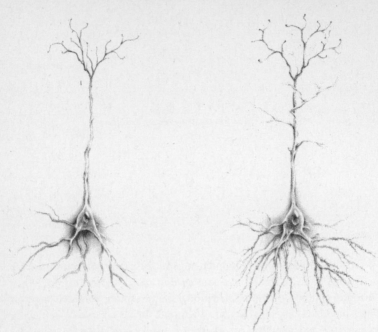

Figure 2. Neuronal Plasticity.
Neuron of a rat raised in a standard environment (left) and an enriched environment (right). Note more dendrites and dendrite spines on the neuron of the rat raised in the enriched environment.
(Adapted with permission from Johansson and Belichenko, 2002.)

In the human brain, networks of neurons are set up to perform particular functions: process information, form and retain memories, navigate in space, recognize familiar faces, parse speech. This is Luria's concept: "Complex behavioral processes," he wrote, "are in fact not localized but are distributed in the brain, and the contribution of each cortical zone [or brain area] to the entire functional system [or neural network] is very specific."

Guided by this knowledge, I came to define a learning dysfunction as follows: an area of the brain that is weaker in functioning than other areas in a network, thereby significantly impairing the learning activities for which that network is responsible. And since a brain area may be involved in multiple networks, a weakness in one area can have wide-reaching impact across many different aspects of learning.

I began creating cognitive exercises for deficit areas before brain imaging made it possible to map their location with any precision. Equipped, thanks to Luria, with an understanding of what I would come to call "the cognitive function" of various areas of the brain, I developed exercises that required students to use their areas of weakness.

When I refer throughout this book to the cognitive function of a brain area, implicit is the principle of interconnected neural networks. Problems in cognitive functioning and learning can occur at many levels: in a brain area, in the connections between areas, or in the network itself.

To change the brain, a program of cognitive treatment needs to deploy what neuroscientists call "activity-dependent neuroplasticity." In 1978 I was using my own intuitive understanding to incorporate this concept into the cognitive exercises I was developing. Simply put, activity-dependent neuroplasticity means that external stimulation that places a demand on the brain over a sustained period of time results in change to the brain. Based on Rosenzweig's research, I took this concept one step further: the activity or exercise has to directly target and stimulate the area being addressed. To enhance the focused stimulation, students are instructed not to call on the support of other, perhaps stronger cognitive areas, which would divert stimulation from the area in question. And attention to the prescribed task must be sustained over a period of time to adequately stimulate the intended part of the brain.

Each exercise, I also understood, must involve what neuroscientists now call "effortful processing." In other words, the task's level of difficulty needs to be carefully calibrated to the individual's level of functioning. If the exercise is too hard or too easy, the brain cannot effectively engage with the task. I thus developed multiple levels for each exercise.

Accuracy, automaticity, and consistency of performance are all essential. If you take 30 minutes to do a task that should take 1 minute or if you can accomplish the task only on a hit-or-miss basis, you are not proficient. For each cognitive exercise, we have determined what constitutes proficient performance. Once a student is proficient, he or she moves on to a more difficult level, one that drives effortful processing again. Reward is built into the program as students achieve proficiency. Rose Kandl, an Arrowsmith teacher in Elizabeth, New Jersey, found an elegant way to explain the process to her students. She had the science teacher at her school connect a lightbulb to a crank; turning the crank would illuminate the lightbulb. Rose told me, "I let students turn the crank fast enough to turn on the lightbulb. I say, 'Now, you have to keep it on. That's what Arrowsmith is. You have to be working at it consistently. If you just turn it on and then you don't work any more, the lightbulb goes out; nothing happens. You want that lightbulb to be going on constantly.' I tell every single child: 'You're going to get out of it what you put into it.'" And if they do this enough, eventually the lightbulb stays lit.

A medical doctor with training in psychiatry coined an apt metaphor to describe life after Arrowsmith, when the lightbulb stays lit. "It's like going out to our place in the country," she says. "I can spend about ten hours in this large building vacuuming cobwebs. They're all over the place. But once I vacuum them all, the place looks like it's supposed to look, so nobody notices a thing. And that's the way it is with my brain. So once it's working like it should work, I take it for granted, and it's like it was always that way."

And once brain function has been augmented, that gain is permanent. My own hypothesis is that once this brain function is in place, normal use offers its own stimulation. I have tracked individuals thirty years removed from the Arrowsmith Program, and there is no loss of function. The changed brain stays changed.

On my long journey to create these many brain exercises, I tested hypotheses and used my intuition and my imagination. Jonas Salk, who developed the first polio vaccine, "became" the immune system as he struggled in the 1950s to find answers to this crippling disease. I remember feeling such affinity when I read that line. In the same way, I "became" the brain.

In his book *Anatomy of Reality: Merging of Intuition and Reason,* Salk wrote: "I do not remember exactly at what point I began to apply this way of examining my experience, but very early in my life I would imagine myself in the position of the object in which I was interested. Later, when I became a scientist, I would picture myself as a virus, or as a cancer cell, for example, and try to sense what it would be like to be either. I would also imagine myself as the immune system, and I would try to reconstruct what I would do as an immune system engaged in combating a virus or cancer cell."

This too was my reality. I had been given a debilitating set of impairments that closed the door on understanding the world through logic and reason, but I had also been given a gift in my ability to intuit the world through the right hemisphere of my brain.

I lived in a world of felt sense rather than one of logic and reason. For me, metaphors were rooted not in language but in my experience of the physical world. When I needed to learn something and thought about diving into material, I physically felt myself diving into what I was learning—as if I were plunging my body into a pool of ideas. When I wrote poetry (something I have done since I was twelve), I became the thing I was writing about. If it was the drop of dew on the petal, I felt its viscosity and wetness, the surface tension, the cohesiveness of the water molecules, the surface energy determining whether the drop would stay contained or flow out over the flower.

People wondered how I managed to understand Luria's sometimes dense and always complex writing on the brain. I relied on that same process. To understand brain function in one area, I became that area, feeling my way into his descriptions of that area and its function, flowing with the world he described—its nature, its particularity.

Creating exercises for various parts of the brain, I would put myself inside that area and try to devise an exercise that would get it working. What activity would stimulate it, and what tools would prove useful? How to begin that exercise? How to make it harder and refine it? With Luria as my guide, I tried to come up with neurological calisthenics.

Creating cognitive exercises and watching what changed in individuals trying them was like conducting naturalistic experiments. I learned more and more about each brain area as it improved. As various cognitive functions came on line, I saw their workings more clearly. Over time I came to understand what happens to the ability to learn when an area of the brain is compromised. Since each networked area in the brain is responsible for a different kind of learning, the impact of a deficit in each area of the brain is also different.

Some students come to us with a striking combination of high and low cognitive functioning. Imagine having a brain that is capable and incapable at the same time. Imagine, for example, having an excellent auditory memory but a severely restricted ability to understand concepts. Given such a brain, gathering information would come effortlessly to you; making sense of it all would not.

Imagine being a gifted inventor and generator of ideas—but owing to a faulty memory, having to reinvent the same thing several times before you actually retain what you have so ingeniously created.

Imagine possessing an excellent visual memory—but a weak ability to produce speech sounds. You thus have a very high level of silent reading and written vocabulary but are confined to simple words in your speech. Your writing would convey to the world one impression of who you are; your speaking, quite another.

All of these scenarios describe someone with a learning dysfunction seen at Arrowsmith School in the past three decades.

Roger Wolcott Sperry, the American neurobiologist who won the Nobel Prize in 1981 for his contributions to neuroscience, noted that there is more complexity in the human brain than in the rest of our physiological makeup put together. He put it this way: "The degree and kind of inherent individuality each of us carries around in his brain—in its surface features, its internal fiber, organization, microstructure, chemistry—would probably make those differences seen in facial features or in fingertip patterns look crude and pale by comparison." Put another way, your brain makes you uniquely you.

Neuroscience holds great promise in that it offers insight into the differences between us and the different ways that each of us thinks, learns, processes information, and responds emotionally—all of which are determined in no small part by the singular makeup of our brains. And with an understanding of the plastic nature of the brain, we can harness this property to positively change its functioning.

What powers this book are the stories of people with brain deficits. Most of these people enrolled in Arrowsmith and overcame their learning disabilities by hard work. Brain work. Some individuals in the book share their struggles with deficits that have yet to be addressed. Every individual interviewed for the book, whether that person chose to be identified or to be given a pseudonym, was eager to tell his or her story. In some cases, identifying details have been changed, but the story itself remains factual.

I selected for this book stories that best illustrate the cognitive areas. I chose young students and I also chose stories of adults to convey what happens when a deficit is left untreated as one matures. The stories of all students who worked on the cognitive exercises have one thing in common: a marked contrast between what was impossible before the deficit was addressed and what is possible now. These are stories of transformation.

My hope is that a greater understanding of how these problems affect people's lives will encourage tolerance, understanding, and, perhaps most important, compassion.

CHAPTER SIX

UNIVERSITY HAZE

1970

I earned my high school diploma, and despite my struggles, there never was any question that I would not go to university. I left high school in Peterborough with a 70 average, and in those days, such a mark was sufficient to pursue a degree.

I chose the University of Guelph, about an hour's drive west of Toronto. My intention was to study nutrition (following in my mother's footsteps), and Guelph was one of the premier institutions for that field. I did manage to get through the first term with marks in the 60 percent range, but I was taking far too many sciences (organic chemistry, physical chemistry, physiology), all of which require cognitive capacities I did not possess. I failed my first research paper, handing in a series of short and unrelated summaries of individual articles rather than integrating the various findings to support a line of reasoning. Everyone else in the class knew what to do. I did not.

Often I would fall into a panic on the bus from Peterborough to Guelph. My only plan was to get off, stand in the field beside the road, and stay there for the rest of my life. I couldn't go forward; I couldn't go backward. There was no place for me in the world. Once again I was a failure. So I switched to child studies.

I had worked at the Unitarian Fellowship in Peterborough with preschoolers. I told my mother and father that once I was in the university environment, I realized this is what I was truly interested in: working with children. And I did enjoy working with children, but that wasn't the reason I switched. I switched because the courses were easier and mostly involved memorizing. Looking back, though, I have to wonder whether destiny was at work in my decision to pursue child studies. Was this a hunt for completeness, a quest to understand what was wrong with me by considering how children learn?

One thing was certain: I began to blossom. I was particularly good in practicums (student teaching or internships) observing children's behavior. We had a laboratory preschool, so we would sit behind one-way mirrors, study children's behavior, and write up our observations. This was probably the first time I was told I had a gift. The work involved nonverbal problem solving and examining nonverbal patterns of social interaction. Jean Piaget (1896–1980) was seen at the time as a preeminent thinker, renowned for studying stages of cognitive development. But no one was making the connection between the observed behavior and the brain.

My struggle to learn continued. I took a year off to travel because I was exhausted from the effort to keep up academically and was not certain I could continue with my studies. My cognitive deficits, though, followed me wherever I went.

On my return, I finished my undergraduate degree (a bachelor of applied science in child studies) and was hired by the University of Guelph. I worked there for two years as head teacher in its preschool laboratory. Privately I was convinced that they hired me because I was a failure: they surely could not allow a graduate such as me to go out and work for someone else because I would reflect badly on the university; ergo, they required more time to raise me up to standard. Most graduates would view being hired by their alma mater as an honor. Not the learning-disabled Barbara Arrowsmith-Young.

By this time, I had become fascinated with the process of learning and why some people could not learn. Important new material on learning disabilities was starting to be published, and I was beginning to explore how differently each child learned.

I eventually decided to attend graduate school at the Ontario Institute for Studies in Education (OISE). I chose to study school psychology, which applies the principles of clinical and educational psychology to diagnose and treat behavioral and learning problems. Here was one more attempt to understand what was working for me, what wasn't, and why I was still struggling.

And those struggles had begun to take their toll. The 1976 Young family Christmas newsletter refers to illness: "Barbara is living with four young friends in Toronto and working on a master's degree in applied psychology. She has not been well the past month and is finding the workload very heavy." In elementary school, there was no actual illness—just a child drowning in her studies. But in graduate school, the cumulative stress from long hours of studying just to stay afloat was having an effect on my physical health.

We now know that stress takes a toll on the immune system, and mine was under siege. Pneumonia was too often the result. I was diagnosed eight years later with endometriosis, a disease thought to involve an immune system deficiency. No longer was the problem confined to the cognitive realm; the stress associated with my learning disability was manifesting itself in my body. I was literally burning myself out.

My professors couldn't conceive of the idea that someone could be gifted and learning disabled at the same time. Convinced by my straight A grades that I could not possibly have learning disabilities, they couldn't imagine that I was spending entire nights in the library to achieve those marks. I would hide when security guards made their preclosing rounds. I needed absolute quiet, no interruptions, and another eight hours with my books: each night alone in the library gave me all that. On average, I was getting four hours of sleep a night.

One day in 1977, I was sitting in the student lounge at OISE when I encountered a fellow student. We began to chat about the brain. This fellow had a master's degree in special education and was pursuing a Ph.D.; he was evidently smart and well read, with an intense interest in the brain. In fact, he was the person who introduced me to the work of Luria. Joshua Cohen had piercing blue eyes that took in the world rather skeptically. He rarely laughed and did not smile easily. What I did not know then was that he would become my future husband.

Joshua and I were married on August 30, 1980. I was twenty-eight and Joshua was thirty-nine. The wedding ceremony was to have taken place in the backyard at my family's home in Peterborough, but ominous thunderclouds (an omen I should have heeded) forced us inside. The local Unitarian minister officiated, and he looked on as Joshua, following the beautiful Jewish custom, wrapped a wineglass in a white cloth napkin, laid it in the fireplace, and stepped on it with his right foot. The custom is explained in various ways. That it symbolizes the destruction of the temple in Jerusalem is the one most often cited, but the meaning that touched me deeply is that a broken glass is forever changed, just as the couple are forever changed by the union. When the glass broke, the small wedding party of family and friends all shouted in unison, "Mazel tov!" Good luck. I wish we had had more of it.

CHAPTER SEVEN

THE FOG IS DISPELLED

1978

Day by day, month by month, as I persisted through the clock exercise I had devised, the metaphorical fog that had enveloped me for almost three decades began to lift. I was quite literally forging new pathways in my brain. The big and little hands of a clock finally began to make sense to me. Then came the defining moment of my life.

I remember in June 1978 excitedly pulling philosophy books from a shelf. I would open one book, read a page, and immediately understand what I was reading, then pull another book from the shelf and repeat, until I was surrounded by a pile of more than one hundred books. Why so many? I wanted to make sure this was not just some random happenstance, so I continued until I had enough evidence that I could read and understand this complex material.

Not long into the clock exercise, I vividly remember watching the *60 Minutes* current affairs program on television and suddenly understanding what was being said as the commentator was speaking. No more having to memorize the dialogue to play it back later. I could follow and understand as the dialogue unfolded. I was watching with my friend Michael, who was always light-years ahead of me in these situations, and I remember his shocked look as I made a comment showing that I understood perfectly what had just been said. This had never happened before. It was as if I had been blind and now I could see.

Long my nemesis, mathematics became fun. Its inner logic, I could now see, had a beauty all its own. I systematically taught myself all of elementary and high school mathematics and thoroughly enjoyed the process. No more math phobia. After barely passing an undergraduate course in statistics, I enrolled in a graduate-level course, understood the concepts, and received an A. And the most amazing thing was that I could understand the material on the first encounter. No more hours and hours poring over books, replaying lectures and conversations. I could understand the first time. Time that had been lost to me all my life was now available to me.

Symbolism in literature was no longer an element to be memorized but an integral part of the tapestry of the written word. My vocabulary expanded and deepened as I began to forge connections and build layer upon layer of meaning.

My writing changed from simple narrative—a chronological listing and summarizing of facts or events—to something else: well-developed and reasoned arguments supporting a central theme.

History, which I once abhorred, became a passion.

I had long loved photography, but with all the explaining in the world, I could not figure out exposures and how to relate F-stops and shutter speeds. After completing the clock exercise, I yet again enrolled in a photography course and was able to understand these relationships and use them to good effect.

I could now understand jokes, especially those involving irony, where the surface meaning is different from the underlying meaning. Understanding the joke requires relating the two—and now I could.

Events in my life that I had experienced as a series of unrelated and random happenings started to have meaning. I underwent an intense and spontaneous period of reviewing my past—like a movie played on a screen. And for the first time, like pieces of a puzzle fitting together, I could make insightful connections between events.

As a result of the cognitive improvement, I had gained clarity. I began to understand that I shared neurological deficits with my parents. I shared with my mother a spatial weakness, which meant, among other things, that we both had a talent for getting lost. My mother piloting the Young family car—in the early days, a beige and white 1958 Chevrolet Biscayne—was always an adventure. But I also inherited her physical energy and task-orientedness and a kind of joy that went with both. "Let's paint the dock!" my mother would say to her children. "It'll be fun!"

I shared with my father a Broca's weakness in the brain, meaning that for both of us spoken language, especially when we were fatigued, was a challenge. It's one reason I needed to close my eyes when I spoke (something I still do) as a way of maintaining focus and a connection between what I was thinking and the words I was saying. But I also inherited my father's creative bent: if there's a problem, let's invent something to fix it.

As a result of my graduate school training, I was familiar with the wide range of standard psychological and educational testing available. I needed a baseline from which I could measure improvement in my brain functioning, and I chose the Miller Analogies Test, which measures verbal reasoning (for example, "mother is to daughter as father is to———"). I had done very poorly on this test when applying for graduate school, and I had been accepted only on the strength of solid references and work experience.

What I found, just several months into my brain exercises, was significant improvement in my ability to understand mathematical concepts, comprehend text, and relate cause to effect. And I was able to do all this not in lag time, as had been the case all my life, but in real time. The ground was no longer shifting beneath me. And my score on the Miller Analogies Test improved by thirty points.

Likewise, I took the arithmetic subtest on the Wide Range Achievement Test some four months into this exercise and found that I had gone up three grades. During all this time, I had not been studying math at all. What had changed was the capacity of my brain to understand concepts and, with it, my facility in math. This was exciting evidence that the exercise I designed had changed an underlying cognitive function. This concept is central to understanding what distinguishes the Arrowsmith approach: cognitive exercises do not teach content or skill in, say, mathematics; the aim is to forge new neural pathways in the brain so that later, when math is taught, number concepts actually make sense.

Today I feel a million miles distant from the frustration and anxiety of that young child, that adolescent, that young woman who was always guessing at meaning. I cannot describe the joy and excitement of being able to really communicate with family, friends, and colleagues, of being able to respond to what they are actually saying in the moment.

I have language now for what ailed me during the first twenty-six years of my life. I know the particular parts of my brain that were not working—their names, their location in the anatomy of the brain—and I know that the exercises I devised to strengthen my brain radically improved my ability to make sense of my universe and be comfortable in it.

What is not so easily shed is the worldview that defined me for so much of my life. Inside me there remains that little girl, frightened by the world and terrified that she does not understand it. Even now, with my better brain and my ability to reason, this deficit's emotional legacy remains.

CHAPTER EIGHT

LOST IN TRANSLATION

"**D**o you want to go to the store before or after lunch?"

That question would have befuddled Zazetsky just as it befuddled me for the first twenty-six years of my life. We both had a symbol relations problem. Owing to this neurological deficit, seemingly simple relational words such as *above* and *below* or *under* and *over* took so long to process that neither of us could keep pace with conversation around us.

If you have a severe case of this deficit you reverse the letters *b* and *d* and *p* and *q* well beyond what is expected developmentally—as I did. You do not grasp that 5 x 3 is the same as 3 x 5, or that 3 - 1 is not the same as 1 - 3. Given strengths in other areas, you may be able to memorize mathematical procedures (such as borrowing in subtraction), but the why of the procedure eludes you. Mathematical concepts such as fractions, percentages, or miles per hour confound you. For example, you would struggle with solving this word problem: If Matt is driving his car at 60 miles an hour, how long before he reaches Toronto, which is 150 miles away?

You might be intelligent, but your symbol relations weakness means you may not comprehend the relationship between yourself, say, and your cousins. Or that your grandparents are your parents' parents. You know the words (*parents, grandparents*), but the concept eludes you.

You can neither get nor tell a joke.

You are stymied by cause and effect, by the rules of grammar, by metaphor. Understanding figurative language poses a major problem. For someone whose grasp of the world is literal and concrete, an abstract phrase such as "He is crazy about that pet" is difficult to comprehend. Likewise, the phrase "off the hook" has two connotations, one literal and one meta-phorical, but appreciating both meanings is almost impossible.

Nor can you grasp the inconsistencies in what someone might be say-ing, leaving you vulnerable to con artists and destructive friendships. You see the world through the cloudy filter of your symbol relations deficit.

I remember one student with a symbol relations weakness. When she was seven and her brother was eleven, she announced to her mother and father that in four years she too would be eleven. She thought that she would catch up to her brother (missing the concept that they would age relative to each other) and that they would both be the same age. No amount of explanation would undo this line of reasoning.

I have had parents with this deficit report to me that they have diffi-culty setting boundaries with their children when their sons and daugh-ters, who happen to have a strength in this area, can outwit them. One mother said that when she announced a new rule, her son would say, with impeccable logic, "What if this happened?" whereupon he would lay out a scenario that made the new rule nonsensical. She had no response.

In another case, a store manager who was responsible for pricing items overcharged when calculating a 30 percent markup on $70. He mistakenly priced the item at $100, figuring that since 30 percent off $100 was $70, the reverse must be the case as well. He saw no flaw in his logic.

With this learning dysfunction, you are prone to what might be con-sidered rigid thinking. You can't easily weigh options and alternatives, so once your mind is made up, it's hard to dislodge you from that position. You may also have trouble understanding and communicating your own thoughts and feelings, which can isolate you socially.

To conceptualize this deficit, hold out your hand with your palm up and your fingers spread out. Imagine that your palm is a concept (an ani-mal or the French Revolution) and your fingers are the facts that must be connected logically to develop or support the concept (dog and cat as types of animals or the various historical circumstances that led to the French Revolution). Now bring your fingers together, the tips touching: this is the synthesis of the elements into a meaningful whole, and insight is born. This cognitive area links seemingly separate pieces of information to one another to develop, support, and validate the concept.

Luria described a problem in the temporal-parietal-occipital area of the brain as a "disturbance of the capacity . . . to combine separate details into a single whole." He called it "semantic aphasia," a loss of meaning.

Neuroscientists refer to this area of the brain as the association area of association areas, since it is uniquely positioned in the brain with connections to those parts of the brain that receive and process information from sight (occipital region), hearing (temporal region), and touch (parietal region). This critical juncture in the brain (which includes the angular gyrus) makes meaningful connections among all those bits of information. In addition, the neurons in this cortical area are multimodal, meaning that they can process different kinds of stimuli (auditory, visual, tactile) simultaneously. For example, the visual image of a cat, the sound of its purr, and the feel of its fur all get integrated in this area to give meaning to what is being perceived.

If there is a significant problem with this part of the brain, received information remains unrelated and disconnected, like the separate fingers of your hand. When the weakness is mild, the learner makes one or two connections, but his hold on the material is tenuous, so there is no depth or stability to the understanding. With so few connections to anchor a concept, the understanding disintegrates over time and comes undone. This is a common refrain from parents of children with this problem: "He understood the material when I taught it to him, but the next day when he wrote the exam, he had lost all comprehension."

This cognitive area makes connections between our seemingly disparate experiences, giving meaning to them and allowing us to build up a data bank of interrelated connections necessary to understand the world around us. The more connections made, the deeper and richer and more stable is one's understanding.

The question children ask, starting around age three, is "Why?" This marks the beginning of an age of discovery as the child quests to understand the world. When the answers cannot be understood, the world is confusing and overwhelming.

The abiding and salient feature of this neurological deficit (and how well I remember this) is an overwhelming sense of uncertainty. As Luria described, it is impossible to ever have any certainty that meaning has been understood correctly.

If ever there was a student lost in cognitive fog and desperate for the clock exercise, it was Zachary. He had five different brain deficits, all of them severe, but the one that caused him the greatest anguish was the same one that most bedeviled me: symbol relations.

A family video of Zachary at age four shows him staring blankly at the camera. "What what what?" he would say over and over again. Or, "Huh huh huh?" So weak was this part of Zachary's brain that he was confused by the "what" of what was going on around him, never progressing to formulate the question "why?" to look for reasons behind the what.

If one can understand the why of events, then one has a certain level of comfort, a feeling of order and control, and a sense that things are not just happening randomly. Zachary's reasoning problem was so severe it left him disconnected from the world. To Zachary, the whole world amounted to just so much noise.

Zachary could not understand most of what people said to him. He could hear the words but could not attach meaning to them, so confusion reigned. Communication was so difficult for him that he didn't dare attempt it with anyone other than his mother or babysitter. Long familiar with Zachary, these two women could understand what he was trying to express and would patiently and repeatedly explain things to him. These two women were his cognitive crutches, and his inability to address others sometimes cost him dearly. When he was four, he needed surgery and was in terrible pain after the operation. For two hours, he said nothing to doctors and nurses attending to him (his mother, Aliza Karoly, was not allowed to see him during that time). He used to bring a phone from home and take it to nursery school, thinking that if need be, he could call his mother at work in order that she speak for him.

On the first day of camp when he was four, says Aliza, "He stood outside the whole time. They didn't know to ask him if he wanted a drink of water. It was ninety degrees. He didn't drink all day. He didn't go to the bathroom all day. He couldn't even ask for himself. He didn't go swimming—because he couldn't ask where the change room was."

Finally, Zachary agreed to go to camp again but only if his babysitter accompanied him, and she would relay to him whatever the camp counselors were saying. Zachary needed a translator to interpret and communicate for him.

Aliza, a personal injury lawyer, first understood that his eccentricity was actually a learning deficit when she was in court dealing with the case of a brain-injured boy. A neuropsychologist hired to assess this child was remarking on how beautiful the boy was and how you could gallop by him on a horse and notice nothing unusual about him. But spend any time with him, and you would begin to see that he needs to be addressed slowly with continuous eye contact; later he might get a job but never manage to keep it. He would need cueing all his life.

"I was sitting in court next to a colleague, I was hearing all this and I said, 'That's Zachary.' I work with brain-injured people all the time, and I had just gotten used to doing that—speaking slowly. It was so normal to me." Zachary was not brain injured, but there was clearly something wrong with his brain.

Aliza was convinced he was bright, but he wasn't communicating; he was miming. He had no friends, save for two neighborhood children and one at school, all of whom made a point of speaking slowly to him. Zachary was anguished by his isolation. For three years, he cried every morning before going to nursery school or kindergarten. In class, he mostly sat by himself in a corner ("his office," the teacher called it) with ear plugs on. Owing in part to an auditory processing problem (diagnosed at school), which in his case, I would say, was the result of his severe symbol relations deficit, Zachary found his energetic and verbal teacher impossible to understand. Zachary had trouble understanding language in any case, and this teacher's style was an additional stressor. The boy would put his head in his hands, bang his head, and say over and over again, "No, no, no."

At home, Zachary had zero interest in cartoons or most television since he couldn't understand dialogue. What he liked were construction shows because he could follow what they were doing by watching the actors. Here was something mechanical, which was and is one of his strengths.

As a result of his neurological problem, there was no order in Zachary's world, so he sought refuge and comfort in things: the sweatpants he always wore, a certain striped shirt, a particular pen, his long hair, sitting in the same seat every day at school.

Zachary was in sensory overload, his world "one great blooming, buzzing confusion"—a phrase penned by nineteenth-century philosopher and psychologist William James to describe the sensibility of a newborn. Zachary had been out of the womb and in the world for several years but could make almost no sense of it.

One result of his sensory overload was that the sight and smells of food repelled Zachary, and he grew painfully thin. Dinner invitations became nightmares. Aliza would warn her hosts that her son might say something wild about the food being served, and, sure enough, he would. "A large part of our culture," Aliza explained, "is just to sit there at a large meal. Zachary would say, 'Your food tastes like smelly old shoes and antiques. Your food tastes like money. Your food tastes like metal.'" Eventually the dinner invitations withered away to those from close family friends.

The symbol relations area of the brain (the juncture of the parietal-temporal-occipital regions) integrates and attaches meaning to information from sensory modalities in order to understand the world. The part of Zachary's brain necessary for processing information coming from those senses and attaching meaning to it was not working properly. He could not attach meaning to language or sensory information. No wonder he shut out the world.

This was the six-year-old boy who came to Arrowsmith School in the fall of 2009. Chris Furtado, one of his teachers, vividly remembers how he was: "Zachary said, 'I hate school, I'm not doing this,' and he would completely and utterly freeze and stop listening to any instructions because he knew he couldn't understand them. He had no ability to reason; there could be no negotiating to even attempt to get anything done."

Zachary had to start working on the clocks exercise, but he couldn't get the connection between the big hand and the little hand. To help Zachary understand the movement of the hands of a clock, Chris told him to get up and hold her hand. "Pretend you're the hour hand and I'm the minute hand," she said. "I'm taller than you, so the minute hand is longer than the hour hand. I said, 'Now I'm walking. What are you doing?' He says, 'I'm walking.' I said, 'Okay, but are you walking beside me or are you following me?' He says, 'I'm following you.' I said, 'That's what happens with the hour hand. The minute hand moves and the hour hand moves more slowly behind it.'" Zachary's deficit was so severe that he could not understand the abstract concept of how the two hands move together until he physically experienced it himself.

Six months into his time at Arrowsmith, Chris and the other teachers began to see a change in Zachary. He became more social. Says Chris, "He was stepping out of the box." The part of Zachary's brain necessary for making connections with the world was so severely impaired it was as if he lived inside a box, the walls isolating him from the world. As that world became less threatening and confusing to him, Zachary could use other people's pencils, a huge change from requiring his own particular pencil and eraser. He needed to feel safe in the world, and now, because he understood the whys, he did. He was starting to reason, to lose some of that anxiety, and to feel good about himself.

"Zachary's comprehension improved," says Chris, "because he got things the first time. There were times in the beginning of the school year where you'd have to explain to him how to do one of the exercises five different ways, and he would still ask the same question because he just didn't understand what you were saying." No longer.

Because Zachary's world had been so confusing and he could not attach reasons to behavior, he could not explain why he did not want to do an activity. He would shut down and give this standard response: "I don't want to do this. It's stupid." What may have looked like a behavior problem was not. "And as this cognitive area improved," Chris says, "he was able to verbalize why he was doing something or why he was feeling the way he was without me cueing him in any way."

Understanding the rules of a game, even one played every day during recess, was at first beyond Zachary. He would rigidly adhere to his incorrect interpretation of certain rules, and no amount of explanation from students or teachers could sway him. Other students would get frustrated and not want to play with him. As he began to process more complex relationships in the clock exercise, we charted his progress during recess. He learned one rule, then the second, and by the fourth rule, his rate of learning had accelerated and he was making connections between various aspects of the game. By spring, Zachary understood all fourteen rules.

When she first saw him, Chris had remarked on how uncharacteristically miserable this six-year-old was. Children that age are often curious about their world and learn through exploration and play. Not Zachary. Eventually he became what Chris calls "a shining star. Everyone in the class adored him."

At home, Aliza saw a similar relaxing, a gradual letting down of his guard. There was no moment of transformation but rather a slow and steady shift as the cognitive exercises strengthened this part of his brain and allowed Zachary to engage with his world. Halfway through his first year at Arrowsmith, Zachary called his father "Daddy" for the first time.

"The biggest change for me," said Aliza, "was that he communicated and had relationships with people other than me. I had been carrying the burden of being his sole communicator. Me and his babysitter."

At home, he demonstrated something he had never shown: a sense of play. "I had a big shopping bag sitting at the front door, a paper bag," Aliza recalls. "And he just decided to pick himself up, sit inside it and start skating himself down the tile floors pretending he was a car, and he'd never ever done anything with any imagination, ever. I couldn't believe it."

In the normal course of development (similar to asking the question, "Why?"), a child uses play to explore and come to understand how the world works. When this cognitive area started to function properly, Zachary started to spontaneously explore through play because his brain could now use play to make meaning of the world around him.

Gone was his sensory overload, his visceral response to loud noises and certain foods. In so many ways and over the course of many months, Zachary had been transformed. Or maybe what we're seeing now is the real Zachary, not that wounded version of himself.

Math teacher Sherri Lane Howie had met him in September, when school began. "He was very reluctant to try new things," she observed, "and became very oppositional with even the suggestion of them. In the twenty years I have been working with children, I have never met someone so young and seemingly so very bright who outwardly displayed such anger and unrest."

Students with this difficulty are often perceived as being rigid due to their inability to consider others' opinions or alternate ways of doing a task. Given the difficulty in understanding all information, they automatically resist trying to grasp and consider something new. Zachary's stubborn resistance was the direct result of a cognitive problem. His teachers all observed that as his reasoning improved, his personality softened, and he was open to receiving and using feedback to modify his behavior.

At the end of that year, the school put on its usual awards presentation. Ian Taylor-Wright, who teaches English at the school, observed that this year's winner had made a powerful first impression on him. The student had walked in that first day of school and matter-of-factly asked why there were so many books in the room. During testing to determine that student's reading level, the pupil assured him it was a waste of time because he didn't understand any of the words. The student never smiled and never got a joke—not even Ian's corny jokes about how superior English is to math. This student took the jokes literally.

The student, of course, was Zachary: winner over seventy-five other students of the English award at the Toronto Arrowsmith School in 2010.

So Zachary is weighing options now, considering alternatives, comparing and contrasting (weighing the pros and cons of this computer system or that book), and making well-informed decisions. This is a fundamental change, one I've seen occur in students over and over again, and a hallmark of improvement in this cognitive function as a result of the clock exercise.

Meet Heather Rayne, a woman in her mid-thirties with a very expressive face. She smiles easily and a lot.

Living with, and overcoming, learning disabilities has given Heather a high level of empathy for anyone struggling to understand, as she struggled for her first twenty-two years.

She has a degree in psychology and is an adult educator now, working as a master trainer for one of the major Canadian banks. In the past, she has taught other bank workers the intricacies of registered retirement savings plans, so numbers and ratios must be right there on her fingertips.

When Heather talks about herself as a teacher, I hear confidence. "I'm very good at explaining things," she tells me. "I still have people coming up to me, people I trained, say, four years ago and they'll say, 'I remember when you taught me this.' Whatever it may be, it'll stick with them, they'll understand it."

Heather now has the ability to understand and explain complex conceptual material in a way that others can grasp. But in her youth, she never could have imagined going to university or teaching, and certainly not teaching anything involving relationships between numbers. Heather had barely managed to graduate from high school, and math had been a major stumbling block. "I was definitely math phobic," she says.

As a student, her enemy—one of them, anyway—was time. Always considered "bright but not achieving her potential," Heather began to realize in high school that homework and in-class assignments were taking her much longer than they took her friends. By the time she had assembled her thoughts to write them down, they had vanished. Heather was a lost student, confused and frustrated. *How,* she thought, *can I be bright and dim at the same time?*

"I would sit in class and feel lost," she recalls. "By the time I processed the question, the rest of the class had given the answer. I would come home from school and tell my parents, 'I don't understand anything going on in class.'"

Testing at Arrowsmith in the fall of 1996 revealed that she had seven different learning deficits, most moderate to mild. Her symbol relations deficit was, in fact, rated moderate.

Distinguishing *nearsighted/farsighted, fiction/nonfiction, East/West,* or the difference among *they're/their/there* were all beyond young Heather. Language and grammar made no sense to her. She was literal in her understanding; nuance in language baffled her.

Any change of plans unnerved Heather. School assignments often confused her and reduced her to tears, and so often was she laughed at for using the wrong word that she shut down both in class and at home. Heather's confidence was shattered.

She had been told she had attention deficit hyperactive disorder (ADHD) and was learning disabled; she was taking Dexedrine, a stimulant used to help her maintain focus.

Then in her early twenties, Heather saw an ad for Arrowsmith in a community newspaper and decided to investigate; she arrived skeptical but full of hope. She had decided she would dedicate one year of her life to cognitive training, and right away she felt the effect of the effort. "My brain would be tired," she recalled. "I was wiped at the end of doing a session."

Heather described to me how living with the symbol relations deficit is like going through life wearing opaque glasses that make everything blurry and confusing. As the brain area acquires the capacity to quickly and accurately process relationships, the glasses come off and the world becomes crystal clear for the first time.

After a year of cognitive exercises, she began a degree in psychology at York University, where she got her first pleasant surprise: a top-ten finish in a class of 112 students. "I was absolutely blown away, flabbergasted," Heather recalls. "I had studied diligently and prepared. But there was always that fear: What if I've completely missed the point? Then it struck me: I've got it, I actually understood."

That year, Heather also passed statistics on her first try. "That amazed me," she said, "because previously I never understood which math formula to apply. I always had to repeat math classes, even in grade nine."

Heather saw other changes while she was at university, all of them welcome. "I realized I didn't have to struggle as much to read something; I didn't have to go back and reread a section. I didn't need to take as much Dexedrine to be able to pay attention. I didn't need to struggle and feel like I sometimes missed the point in a social conversation. I wasn't quite as shy as a result. I still sometimes think of myself as shy, but overall most people who know me would laugh that off right away."

She now works in the bank's fraud management department. The same woman who couldn't catch the drift now helps to catch the grifters. And the Dexedrine is a thing of the past. What had looked like ADHD was actually attentional drift due to the learning disabilities. Once they were addressed, lack of focus was no longer an issue.

Many students come to Arrowsmith with a diagnosis of attention deficit disorder (ADD) or attention deficit hyperactivity disorder (ADHD). Over the years I have come to put attentional problems into one of four categories. The first involves attentional difficulties due to emotional factors; the second involves subcortical problems (those regions of the brain below the cortex involved in regulating states of arousal and attention). Neither can be addressed by the Arrowsmith Program. In the third category, a number of cognitive deficits conspire to make it hard for the student to sustain attention when the task at hand requires these areas to function well. The student tires and attention wanders since much greater effort is required to perform the task that recruits these underperforming areas of the brain. This was the case for Heather Rayne. The fourth category involves a weakness in either the left or right prefrontal cortex (more on these areas in Chapters 9 and 11). A critical feature of these two areas is the sustained attention to task, so a deficit in either area will naturally result in attentional problems.

As we address and strengthen the functioning of the cognitive areas in these last two categories, students are able to come off medication and sustain focus naturally because learning is no longer difficult and because the brain now has an enhanced ability to regulate attention.

Brocha Neuman, an Arrowsmith teacher in a boy's yeshiva (an Orthodox Jewish school), sees improvements in her students' ability to understand even that notoriously difficult sacred text, the Gemara. The Gemara is part of the Talmud, and its style is to weigh many sides of an argument, just as a clock in the Arrowsmith exercise can have multiple facets. Students who work on the clock exercise become quicker at grasping and juggling the multiple related ideas necessary to put forth a reasoned argument.

A lawyer whose symbol relations deficit was playing havoc with her professional and home life describes what motivated her to seek out and stay committed to the clocks cognitive program despite her incredibly busy life. "When I prepare for appeals court," she said, "I read dozens of cases all saying the same thing. One case would do—if I could only understand the reasoning behind it. But I don't. I keep thinking that if I get enough details, I will understand what it all means, but I can't follow them. I am always guessing at what they mean. It's agony. Every time the phone rings, I worry that I've screwed something up." She was afraid she would misinterpret a key point of law and not only expose the client to risk but expose herself to litigation from the client.

Seven months into the program, she reported the following changes at work: "My mind doesn't shut off. It works on everything. I can understand and consider multiple angles when presented with a problem. I know the reasoning has been stepped up. I am deadly on cross-examination. I love the puzzles in litigation now. It is a tremendous challenge to set up my case and work it through. It's really exciting now that I have a brain to start working with. I am now reading philosophical texts and understanding them."

At home, the normal squabbles between husband and wife had begun to tilt in her favor. "When we fight now, unfortunately, it's not a fair fight. I understand where he's coming from and can anticipate his arguments. He wants to work on this cognitive exercise to step up his functioning so he can take me on again."

She was also observing a change in her relations with her children: "That's the nicest part, starting to understand my children. I'm getting along great with the children. I have long talks with my family, analyzing their problems."

With her symbol relations capacity significantly changed, this woman, this lawyer, this mother was finally able to say, "I understand."

CHAPTER NINE

HITTING THE WALL

L et us enter the world of someone with a symbolic thinking deficit. With this deficit, the frontal lobe, and, more precisely, the prefrontal cortex on the left side of the brain, is underperforming.

Imagine a wall before you, a wall that extends in either direction as far as the eye can see. Your task is to get past that wall and continue on your journey. The obvious questions are these: Can the wall be vaulted or penetrated? Might one succeed in tunneling underneath the wall? You survey it, note its height and thickness, and wonder whether a rope and climbing gear, a chisel, or dynamite might figure in the solution. The wall is a problem that requires problem solving.

Now imagine that the wall provokes no thinking at all but only mental paralysis. This is what it's like to have this neurological deficit. A symbolic thinking deficit robs you of mental initiative and leaves you distractible, disorganized, and unable to plan and set goals.

In 1978, I had been reading one of Luria's books and specifically a chapter titled "Disturbances of Higher Cortical Functions with Lesions of the Frontal Region." I was thinking that I had never worked with anyone beset by a problem in this area—the prefrontal cortex.

And in April of that year in came Johanna Vandermeer, a twenty-two-year-old bank teller. More than thirty years later, hers remains the classic case of this neurological deficit. It was Johanna who came up with the metaphor of the wall.

"As a child," she later wrote, "I was taken from psychiatrist to psychiatrist—all of whom told my mother that there was definitely something wrong with her child but they didn't know exactly what it was or how it could be fixed.

"In high school," Johanna continued, "I was the communication secretary in the school's audiovisual department. The director of this center taught all my courses. Because I took part in this audiovisual center, I didn't have to go to many of my classes. At the end of my graduating year, I was asked by my teacher what grade I wanted, an A or a B. I took the B because I didn't think my parents would believe I got an A. This is how I graduated from high school."

Throughout her schooling, Johanna would answer without thinking, not check her work before handing it in, and neither reflect on nor learn from her mistakes. These are all features of a symbolic thinking deficit. She could do mechanical math procedures, but she could not problem-solve or apply the mechanics appropriately. She had huge difficulty developing a theme or idea in her essays and could not separate salient facts from trivial ones. For Johanna, every piece of information had equal weight, so determining the main idea of a discussion or written material was impossible.

"I could never make a decision," she recalls. "In 1976, I went to university. I lasted three months. Four different living quarters, all the wrong friends, tears and trouble. The last straw was getting F's and D's at school. I felt stupid and useless."

Johanna caused her parents enormous distress with her lack of judgment and inability to anticipate risk or foresee consequences. A roommate, for example, told her a hard-luck story about his financial problems, and she gave him all her furniture. Her parents were constantly having to rescue her from desperate situations such as this. Everything Johanna described fit with what I had been reading in Luria.

The prefrontal cortex is critical to effective and efficient functioning. Choose your metaphor. If you follow business news, think of the prefrontal cortex as the brain's chief executive officer. If you're a fan of football, imagine the prefrontal cortex as the team's coach. Or if military history is your passion, cast the prefrontal cortex as Napoleon Bonaparte or the Duke of Wellington devising strategy and issuing commands as the battle unfolds. Imagine a company without a CEO, a football team without a coach, an army without a general.

The prefrontal cortex helps us to formulate goals and intentions and pursue them. People with problems in this part of the brain are often seen as irresponsible, flighty, impulsive, and indifferent to their own problems. These are not "personality traits" in some psychological sense, as is sometimes assumed, but the result of a neurological deficit. Brain exercises that tackle this cognitive disability make these individuals more self-directed, less dependent, and more capable of strategizing and adhering to their goals.

Luria defined thinking as "a special form of cognitive activity arising when a person is confronted with a problem, or in other words when a goal is placed before a subject in certain conditions and no prepared solution to the problem is available." He distinguished intellectual activity from simple trial-and-error behavior and broke the thinking process into three clearly defined phases. "The subject," wrote Luria, "who tries to solve a problem in an organized manner must first discover its conditions, and pick out the essential elements; he creates a hypothesis or general plan (strategy) for solving the problem and selects the methods (operations) by which this may be done." For Luria, this preliminary phase of the thinking process is also the most complex. Then comes the second phase, "which is executive in character and consists of the application of a series of operations leading to the required answer." In the last phase, problem and answer are compared. If the problem has been fully solved, wrote Luria, the process is complete; if it has not, "a state of 'discord' arises and the process of seeking the required solution continues."

Unable to explore fully all elements of a problem, individuals with this deficit tend to get stuck at phase 1 of the thinking process, and any solution generated (a huge challenge in itself) will be limited in its effectiveness. These individuals rarely get to phase 3.

"What should you do if you are in a lecture hall and are the first person to see smoke and fire?" Johanna was asked during her assessment. Johanna replied that she knows what she has been *told* to do (inform someone in authority). What she would actually do, she said, would be to tell the person next to her. She would react without thinking to the immediate aspects of the situation—even when her own life was theoretically in danger. Acting was easier than thinking. Best, Johanna told me, was to have time beforehand to discuss with someone a course of action and then follow it. In practice, that rarely happened because in the moment, she would act, not think.

Another feature of the symbolic thinking deficit revealed itself as we tested her. Discussing a topic (what Johanna needed to learn in her job, for instance), she would be derailed by an unrelated thought—for example, why could she never finish university?

A family connection had helped her get her current job, at a bank, but the employer had recently contemplated letting Johanna go. She could learn procedures but lacked the judgment to know when to apply them. Introducing some new element, one that meant modifying the procedure, paralyzed Johanna. She was not seeing her mistakes, and when they were pointed out to her, she could not figure out how to correct them. When she made an error, she could sense that people were upset with her, but she could not change her behavior in order to avoid the same mistake.

Johanna's mother helped her devise rules that allowed her to perform in a few different situations. But Johanna was never promoted beyond a very junior position. She brought no mental initiative to the job: she could do exactly what was asked of her, but no more. Unable to reflect on her performance, evaluate it, and determine how she could improve, she was immobilized. She had trouble organizing her time. She was easily distracted and could never finish what she started. Each task was its own wall, and she had no means to get past any of them, big or small.

"I see people at work learning things," she told me then, "and I feel that I am as smart as them, but I can't figure out how to do what they are doing. It is very frustrating. I see them being promoted, and I am stuck."

Johanna told me that she'd been accused of lacking motivation. Showing real insight, she said, "It is not motivation I lack, but a sense of direction, an ability to plan forward. I want to do things, but I can't figure out how to go about doing them."

Johanna fiercely wanted to be independent, but she kept running into trouble and needing help from her parents and friends. She could not do the simple things that most of us take for granted. Upset because her face was continuously breaking out, she would try to hide the problem. And then a friend suggested she see a dermatologist. Seeking medical help had not occurred to her. She wanted to lose weight and began to succeed only when a friend gave her strategies to follow. The right hemisphere of her brain worked fine—she was appropriately upset or emotionally responsive to situations—but she had no answers to the myriad problems that arise in the course of a day.

Johanna had been tested at the YMCA Vocational Counselling Centre, where a counselor named Reg Bundy determined that she had a learning disability specifically linked to problem solving. On a test to evaluate critical thinking (her ability to interpret material and recognize themes and overall ideas), she scored very poorly. Reg Bundy suggested she come to Arrowsmith.

I set out to create an exercise for Johanna that would stimulate the left prefrontal cortex—the part of her brain, I was convinced, that desperately needed stimulation. Here is Johanna describing the exercise: "The exercise I found exceptionally helpful was reading fables for hours and finding what the real meaning was. What was interesting is how the stories pertained to real life." The purpose of a fable or parable is to teach a moral or life lesson. Someone with a problem in the symbolic thinking area of the brain often gets caught up in the details of a story and cannot easily discern its message. They cannot separate the kernel of truth from the surface features of the narrative. This parallels their approach to life. When this area is working well, it enables us to dig below the surface to see the deeper intention. This Arrowsmith exercise uses fables to work the brain so it's better equipped to separate wheat from chaff.

Several months into the program, Johanna started to notice improvement. At first, she wasn't enamored of the change. I would later see this same phenomenon in other students. Flawed though Johanna's thinking process was (indeed, there was little process at all), that approach had come to be tied up with her identity and self-image, and now both were undergoing a transformation.

She came in one day looking perplexed and asked to speak with me. Johanna had observed that she was becoming more objective about situations and no longer reacted impulsively. Now she would step back and analyze a circumstance to figure out the appropriate response. Johanna felt as if an internal stop sign had been installed and was preventing her from acting without thinking first. Johanna was not sure she liked this new sense of responsibility and feared she was losing her spontaneity. Several months later, though, when a serious family crisis occurred, she responded in a more considered manner than any of her relatives, and for the first time, they came to her for advice and support. As Johanna improved over the next year and a half, her employer started to notice. She was sent to take courses and promoted to a supervisory position.

Two years later she wrote to me: "I always wanted to learn everything and get ahead, but I couldn't before. I have now completed courses on my own at the University of Toronto. I took a course in effective persuasion at Ryerson and received a B. I then took marketing, in which I got an A. In the latter course, I had to do case studies on how to build a company. I had to think these through—and guess what? I did, and got As and Bs."

I spoke to Johanna in September 2010, thirty-two years after we first met. She's married and has a family and a flourishing business. Johanna's ability to think strategically allowed her to develop contacts and networks to get work in her field, advance her career, and start her own business. That feat would have been impossible, she says, without the cognitive work that brought the wall tumbling down, to her immense relief.

Only in the last several decades has neuroscience become aware of how critically important the prefrontal cortex is to overall cognition. When it was surgically removed or separated from the rest of the cortex (through, say, frontal lobotomy), neurologists found that speech, memory, motor skills, and scores on standard tests of intelligence showed no change. But the right questions were not being asked, so the actual functions of this area remained unknown.

Two Canadian neurosurgeons, Wilder Penfield and Joseph Evans, noted in 1935 that "large amputations of the frontal lobes . . . produce surprisingly little disturbance of function which can be detected by ordinary methods of examination. . . . Neurologically . . . and by the ordinary psychometric tests, each [case] would have to be judged normal." The two men do, however, intimate that something did change, but whatever it was could not be measured. "[This patient] has lost something," they wrote, "that psychometric examination does not evaluate. He has lost initiative; not all of it, but much of it."

We now know that this part of the brain comprises one-third of the neocortex, includes areas that are unique to humans, continues to develop until at least the early twenties, is extensively interconnected with much of the brain (including subcortical areas), and is vital to overall intellectual functioning.

The prefrontal cortex is the anterior part of the frontal lobe, in front of the motor and premotor areas. It has been characterized as the "executive" of the brain and often referred to as carrying out "executive functions" necessary for planning, decision making, evaluating behavior in terms of goals, sustaining attention, suppressing impulses, inhibiting inappropriate responses, evaluating behavior in terms of past actions and future consequences, and delaying immediate gratification for more long-term reward.

As a student of child studies at university, I remember watching a video of an interesting experiment that pointed (though nobody knew it at the time) to the importance of the prefrontal cortex to later life success. In 1972, psychologist Walter Mischel conducted what is now known as the Stanford marshmallow experiment. Four- to six-year-old children were presented with a marshmallow and told they could eat it now or wait fifteen minutes, when they would receive two marshmallows.

The responses were recorded in terms of length of time the children could delay gratification. At a follow-up in 1988, the young adults who as children had delayed gratification were described as significantly more competent than those who had been unable to do so. The ability to delay gratification was found to significantly correlate with higher Scholastic Aptitude Test (SAT) scores.

What was actually being measured in this simple experiment was what underlies the ability to delay gratification: the functioning of the prefrontal cortex, which is critically important to success in life.

The symbolic thinking area allows us to consider everything required to achieve a goal: explore all possible paths and their consequences, choose the best strategy while constantly weighing that strategy against the intended goal, and suppressing impulses and ignoring distractions.

Think about that analogy used earlier: separating wheat from chaff. We are all faced with internal impulses and external distractions on a day-to-day, if not minute-by-minute, basis. Some people act on those impulses and distractions, and some don't. Why? A strong symbolic thinking capacity allows us to use good judgment, which includes ignoring impulses with likely undesirable outcomes.

When an individual is described as ambitious, focused, or driven, that person likely has a strength in this cognitive area; he or she can suppress or ignore impulses and distractions and maintain focus on the goal at hand. Conversely, an individual with a symbolic thinking deficit will lack self-directed organization. This individual has difficulty creating a goal-oriented plan: deciding what's critical and how to start. In many cases, parents and teachers try to create structure for these individuals. Individuals with a mild or moderate difficulty may be able to follow a structure laid out for them, but because no structure can cover every situation, they have limited success. When problems arise, these individuals are unable to make midcourse corrections.

Individuals with a severe symbolic thinking deficit are at an even greater disadvantage: they cannot maintain sufficient focus to carry out a plan, even one designed specifically for them by someone else, so they get sidetracked, lose focus, and feel overwhelmed. When distracted, they give the distraction equal value and attention and cannot easily shift their focus back to the task at hand. They simply don't know what's relevant to accomplishing their goal.

Nor can they multitask, which requires an ability to shift quickly and appropriately between tasks as each one takes precedence at different points in time throughout the day. Due to these difficulties with attention, these students are frequently diagnosed as having attention deficit hyperactivity disorder.

A term used to describe this region's role is *cognitive control,* or the active maintenance of behavior in the service of achieving goals. According to two MIT neuroscientists, Earl Miller and Jonathan Cohen, this part of the brain provides "topdown bias signals to other parts of the brain that guide the flow of activity along the pathways needed to perform a task." The authors suggest that this area controls many aspects of human cognition.

When presented with a problem to solve, the symbolic thinking area of the brain is the recruiter, calling on the other brain areas necessary to perform the task and maintaining activation until the task is completed. When symbolic thinking is weak, activation is not effectively maintained, and there's a ripple effect: an individual's ability to use strengths in other areas is effectively limited.

Working memory, a term first used in the 1960s, refers to the capacity to hold and manipulate information in one's mind for brief periods of time. A simple example is to repeat seven numbers forward, wait thirty seconds, and then reverse the order of the digits. Researchers have found that the ability to retain and manipulate information depends on the left prefrontal cortex, which recruits multiple cortical areas and different neural networks depending on how the information is presented (visually or auditorily, for example) and its nature (language, numbers, pictures).

Essential to any working memory task, the left prefrontal cortex sustains directed attention to what is to be remembered, suppresses responses to irrelevant stimuli, and modulates and maintains activation of any neural networks called on. The other cortical areas recruited for the working memory task are also important in successfully completing the task. A person who has a deficit in the areas responsible for number sense will find a working memory task with numbers to be difficult, even if that person has strong functioning in the prefrontal cortex.

A study in England in 2009 suggests that working memory impairments in children are associated with poor learning outcomes and constitute a significant risk factor for educational underachievement. Children aged seven to eleven with identified learning disabilities were assessed on measures of working memory, IQ, and learning. When the children were retested two years later, working memory, *not* IQ, best predicted learning outcomes. If what is predominantly being measured in this study is prefrontal cortex functioning, which plays a significant role in working memory tasks, then this is consistent with what I have seen with individuals over the years. Strong functioning here is a predictor of later academic and life success. Weak functioning in this area is related to poor academic and job performance.

I recall a twelve-year-old student with average intelligence but whose severe weakness in both the left and right prefrontal cortexes left her as compliant as a young child—so compliant that other children would toy with her and order her to stand and sit on command or to stay in the schoolyard long after recess was over or to surrender her Nintendo game. Her neurological weaknesses had robbed her of her ability to evaluate a command and decide whether it should be obeyed. She addressed her problem areas and eventually was able to say no.

Another student, an accountant, had taken literally our instruction to "fill in the blanks" with a pencil during a timed test. Despite our repeated reminders not to, he was wasting precious minutes ensuring that each and every circle was completely blacked out. He obeyed that first command and, owing to a symbolic thinking deficit, could not see past it.

Gabriela Sanders has big, dark, expressive eyes, and she laughs easily. If she teared up several times during our conversation, that was because she was reliving grim episodes from her past. Now a health care practitioner in Toronto, Gabriela was born with learning disorders and failed tenth-grade math three times. The impact of the learning disability only got worse throughout high school.

"School was really challenging," Gabriela recalls, describing the classic hardships that I associate with a symbolic thinking weakness. "I never knew how to focus. I could never sit down and do homework. I didn't hand in assignments, wasn't able to do a test because I couldn't study for it, so I would sit and wait for the time to go by. It was seen as defiance. The conclusion that I came to was this: if I can't sit here and do work, it must mean I have a real problem and I am probably a really bad kid. So I am going to act out and prove to them that they are right—that I am a bad kid and watch how bad I can be."

"I remember being overwhelmed once I hit high school," says Gabriela. "It was all too much, just all too much."

Like a child riding a bike with training wheels, Gabriela did fine—until the wheels were removed and she was expected to find her own balance. The earlier grades are about learning to read and write, and Gabriela had strengths in areas that enabled her to use phonics to learn to read and also retain sight words visually.

Those demands of the early grades—teacher-directed rote and skill learning—don't require much independent thinking. Grades 7 and 8 begin to place demands on independent thinking, along with the expectation that students start to learn on their own. Gabriela's symbolic thinking impairment was rated as mild, but it's a measure of the critical importance of this part of the brain that even a mild weakness can have a significant impact. Aided by much support, a strong visual memory, and a number of other areas of strength, Gabriela got through high school and college and later pursued a career in health care. Still, she acutely felt she was limited in what she could do.

Before embarking on a program at the Toronto Arrowsmith School in 2009 at the age of fifty-three, Gabriela was tested. None of the findings surprised her.

Within months of beginning the exercises, Gabriela started to see change. Initially, she suspected a placebo effect, but her husband was the one who first noticed that she was better, for example, at navigating while driving. He excels at this, and she had always deferred to him. More present to the world now and able to pay more attention, she challenged him.

Gabriela was also getting better at remembering where she had put things; she could now stay focused long enough to construct mechanisms by which to remember. More important, she was clearer in her speech and thinking; she could now organize her thoughts to speak coherently without getting distracted by irrelevant thoughts. She is now able to map out in her head the structure of, say, a health-related workshop. Gabriela had always relied on her husband to help her with the structure of a research report. Like many other individuals with this deficit, in the past she had taken a cookie-cutter approach, so all her writing followed one model unless someone helped her. She could not, on her own, generate alternative ways of writing. No longer. And she now quickly and with less anxiety figures out the core idea of what she reads.

"Even my therapy," she says, "has progressed as a result of the cognitive changes. I could never really sort it all out before. It was just like a big hodgepodge of thoughts, emotions. What's a thought? What's an emotion? My thinking is clearer. I can see the essence of the problem at hand. It's such a relief. I don't feel stupid. Now I can read the newspaper and say

'That's a good point' or, 'That's a crock.'" When this deficit was in place, everything that Gabriela read had equal weight. She could not discern which pieces of information were critical and which were irrelevant.

As cognitive strengths develop (and as the lawyer described in the previous chapter), even something as apparently set in stone as the balance of power in a decades-old relationship can undergo change. Gabriela describes a trip to Australia during which she and her husband had a long discussion on a favorite topic: theology. "It was interesting," Gabriela recalls. "He finally threw in the towel. I was able to keep him on track and going back to the original point. When he saw that he wasn't able to influence me in a particular way, he started taking it off to another place, and I was able to say, 'Hold on a second. That's not what we were talking about; we were talking about this.' We had little chuckles in there. At the end of it he said, 'Wow, that was really good.' It was mentally challenging, but I was able to stay with it. I could never do that before."

When asked about the Arrowsmith Program by the parents of young people struggling with learning difficulties, Gabriela talks about "short-term pain for long-term gain." As a health care professional dealing with addictions, she well knows where ignoring a learning problem can lead. As for adults such as herself, Gabriela now understands that it's never too late to address a learning problem. The end result, in her case, is an option she never thought possible.

"I'm thinking about studying anthropology," she told me before leaving. "I would have never thought of that before. My world has really opened up."

The file on Stuart Davies is ample and marked by a great many optimistic notes (his journals describe how "the gridlock" in his brain feels like it's been "surgically removed"). But I have to say that when he left my office in October 2010 after a lengthy chat, what I felt most was a great sadness.

We have known Stuart since 2002, and I don't know that we've ever had a gentler soul in our midst. He had come to Canada as a teenager in the mid-1970s, trained as an accountant, and worked for many years as a controller for small firms. What he did not know is that he had half a dozen cognitive deficits, and the collateral damage from them is sobering: five lost jobs (plus two where he worked day and night to stay afloat) and a shattered marriage.

When Stuart sat before me that fall, the smile on his face was constant even as he described his lifelong struggle. As he noted in the journal he kept while attending Arrowsmith School in 2004, "I have suffered too long."

Accountants need to anticipate what's required and to problem-solve as necessary. And since one of Stuart's neurological deficits was symbolic thinking, he was saddled with a passivity that hindered his ability to prioritize and strategize. He wasn't lazy; Stuart typically worked longer hours than anyone around him. "Work harder, not smarter," was the regimen his learning disability forced him to adopt.

I found it painful reading a series of e-mails sent to Stuart in 2001 by his superior at that time. The subject was a performance review, and although the empathy that Stuart's boss felt for him was evident, so was the frustration. Stuart's mistakes were grave and many: deposits not made in time, input errors, disorganization (an office so disorderly it inspired no confidence in its tenant).

Stuart could handle the rudimentary aspects of his job, such as maintaining the payroll account, and he was actually quite a good accountant once a structure or procedure was imposed on him. But anything beyond that would overwhelm him. He was like a photographer with only a standard lens at his disposal but who also needed a wide-angle lens to get the big picture. Stuart's employer simply assumed that his accountant would stay current on tax and accounting changes introduced almost annually by Revenue Canada and that the books would be impeccable were they ever to be audited.

But, no, Stuart was consumed with routine day-to-day accounting. The initiative to think past that was simply not in him. Stuart's symbolic thinking deficit contributed mightily to this failure to anticipate.

A measure of Stuart's desperation is that when he lost the job at the firm, he and his family decided that his severance money would be best spent dealing with his cognitive weaknesses. For two years, he was a full-time student at Arrowsmith. Stuart made great strides, and mentors tell him he's more focused, more coherent, less inclined to jump into a project headfirst. Stuart is so much more capable and in so many ways. But he hasn't "got back on the horse," as he put it. He no longer takes on big jobs but chooses instead to work under a supervising accountant and on a freelance basis. He worries that taking on more responsibility would mean a return to the anxiety that almost crushed him. These junior positions, though, offer less money, which introduces a different kind of stress into his life. Stuart had not addressed his learning difficulties until the age of forty. He was able to strengthen the key areas to the point where he can now problem-solve and take initiative, but the years of anxiety have taken their toll and left an indelible mark.

If writing comes easily to you, it might be hard for you to imagine the struggle of someone with a symbolic thinking deficit trying to stay on topic. I once asked students to write a composition titled, "Should Students Be Allowed to Bring Pets to School?" One essay stood out as a clear demonstration of the symbolic thinking problem. I've put in italics the places where the wandering off topic begins:

> One good idea to bringing a pet to school is because you might be lonely. It is *best when you bring a pet to an old folks place or retirement home, it's a breath of fresh air; reminding them when they had a pet during their younger years.* Back to the pet to school topic. It might get stepped on, people might be allergic to them or they don't like the pet. On the other hand, people have enjoyed the companionship with their pets. *And it is hard to let go when a loved one loses a job and can't afford to feed the pet or take care of it. Or the family moves away and they leave the pet abandoned or in a shelter but sometimes the pet makes it back to the family as it is determined to do so.*

The student clearly tried to remain on topic, but even after reminding himself to return to the question, he could not help but go astray and write about irrelevancies. Because this area of the brain is critical in keeping us focused on the overall goal or demand of the task, a problem here leads to loss of relevant focus and wandering off in multiple directions.

Karen Imrie stands alone in her apartment, knowing that a friend is coming over shortly for a visit. But the thought of cleaning her place is so overwhelming for her that she is paralyzed, literally frozen in her tracks. *Where do I begin?* is the huge question, for which she has no answer. From her vantage, she can see dishes piled up in the kitchen, papers on her coffee table, and clothes covering her bedroom floor. In every room are items that belong in another room. All these details have driven Karen to distraction.

On the wall is an organizational tool she has set up for herself but forgets to use. What might look like a memory problem is really a lack of mental staying power to follow through on a plan. Karen can't complete one task before starting another, leading to paralysis and to what some might consider a personality trait: procrastination.

An individual with a symbolic thinking deficit never becomes skilled at the most efficient method of accomplishing, say, domestic chores. Given half a dozen tasks in the kitchen and laundry room, it would make most sense to do all the chores in one room before moving on to the next. But people with this deficit waste time going back and forth, and there's a good chance that distraction will derail everything.

Karen's inability to accomplish the seemingly simple task of cleaning her apartment had nothing to do with her intelligence. When tested in the fall of 2007, she scored in the 96th percentile in basic intellectual ability (average is 25 to 75), and in mathematics she scored in the 97th percentile. There was no doubt that Karen was intelligent, but it was clear to me that one part of that extraordinary brain, the left prefrontal cortex, was malfunctioning.

She went to university, but she would enroll one semester, then drop out the next. Ever since primary school, she had had trouble deciphering the main point. For most of her life, the question that always immobilized her was, *Where do I start?*

"Sometimes it was difficult," she recalls, "just to stay with the core of any assignment. I would want to add lots and lots of details, not knowing the difference between what was relevant and what wasn't." Eventually she went to Japan and taught English.

Fifteen years later, immediately following her second attempt at university studies, she sought work outside an office. "I didn't have the mental capacity," she explained. "I had no idea what people did in offices. I'm very practical. I needed hands-on to have it make sense. I always enjoyed working with my hands," she said. "I was the builder. I had three brothers, and I was the one who fixed the washing machine and the toilet."

Our testing showed that Karen's mechanical reasoning was at the 97th percentile, a mechanical aptitude that landed her nontraditional roles in the workforce. Karen drove 10-ton trucks and spent a year as a mobile crane operator. She also worked on planes in an aircraft factory and as a carpenter. These were unionized jobs that paid well, but as she neared retirement Karen found herself beset by an old question: *Given my intelligence, given that the rest of my family had gone to university, why not me?* She thought back to her struggles in academe.

By her third year at the University of Toronto (during her second attempt at university), Karen recalled, "I was ready to have a lobotomy, I was ready to have shock treatment because the frustration level was sky high. I wanted a change in the way my brain functioned, so it would be able to do what it was supposed to do."

Karen came across a book on attention deficit disorder called *You Mean I'm Not Lazy, Stupid or Crazy?* She considered but quickly dismissed the possibility that she was *stupid* or *crazy*, but she found herself pausing at *lazy*. How else to explain the uncommonly messy apartment? Unfortunately, individuals with learning challenges are frequently thought of as indolent. They are obviously intelligent and yet can't structure themselves to complete a project at hand.

Karen developed two strategies for dealing with her apartment: she avoided having friends to her place, or she would pile accumulated papers and other objects into cardboard boxes and tuck them away in the nether regions of her apartment, creating at least the veneer of tidiness. At one point, Karen had twenty-six such boxes.

In the Arrowsmith classroom, Karen immersed herself in the "wheat-from-chaff" narrative exercise meant to combat the symbolic thinking weakness. Luria had suggested that the preliminary, and most complex, part of the thinking process involved picking out the essential elements of the problem at hand. Before Karen could even begin to solve the problem, she had to figure out, as Luria said, "its conditions." Which were the salient details, and which were extraneous?

Day after day, Karen pored over text—simple narratives at first, then more complex. She was searching for solutions to a range of problems presented in the material, generating solutions and discarding unworkable ones before finally embracing ones that worked. All this effort stimulated Karen's prefrontal cortex. Over time and in all aspects of her life, she started to think independently and to problem-solve. No one was more astonished than Karen herself.

"This," she says, "is just what happened. Literally within a few months, all of a sudden I realized my apartment was neat. It had stayed that way for three weeks. I wrote a note saying, 'What's happening to me?' About a year later, a friend drove me home, and I had something in the apartment that would have been useful to her. And I said, 'Well, listen, come up to my apartment and I'll look for it and I can give it to you.' What am I saying? A year ago this never would have happened. To invite somebody—without three days' preparation or clean-up."

The other change that Karen noticed was her ability to stay on topic. A friend used to tease her: he would keep tabs on how many times her mind wandered from what was being discussed during dinner, and his tally might reach seventeen. But after one year of Karen doing cognitive exercises, the same friend was complimenting her. "Karen," he told her recently, "this whole dinner you haven't gone off topic once." He had defined her by her Ping-Pong mind, and now he needed a new definition. These days Karen is not distracted by the radio or other conversations, and when she tells a story, she can do so succinctly. And she is down to three boxes of clutter in her apartment now.

When I last spoke with her Karen was just shy of sixty-two, not much older than me. Her friends and mine now joke about becoming forgetful—having to retrace steps to locate missing keys, glasses, and purse.

Karen sees some wisdom in people her age taking cognitive courses. "I would challenge you," Karen says, "to find somebody who doesn't have an area of weakness."

June Winters came to us in the mid-1980s with a classic symbolic thinking profile and did cognitive exercises over the course of several years. Her husband, Robert Winters, a physics professor at a university in Europe, later sent me a note. His letter, dated March 28, 1990, astutely chronicles his wife's experience:

> Dear Barbara,
>
> For many years, my wife June, who is now fifty, had been unhappy with her inability to succeed in a job, and in many ways had found working something of a revolving door. She felt "stupid," and although she scored in the upper 120's on standard intelligence tests, she was unable to function in an office environment. She just kept "forgetting," or "misunderstanding" instructions and situations.
>
> The problem was not a simple one of poor memory, since June was able to remember facts, names, and incidents better than most other people and has an amazing vocabulary. Rather it was an inability to deal with unexpected situations as they occurred. When situations were structured, June did very well. When something unexpected happened, June was at a real loss.
>
> In our personal life, I avoided having her go to the bank on any but the most routine errands, since I was sure that the result would be that either she would have to do it over, or

that I would have to undo what she had done. In company, she avoided involvement in discussions since she felt that she had little to add to the conversation.

It upset June that our children were more likely to come to me with a problem than to her, as experience had taught them that she couldn't help them find a solution.

She had long known that something was wrong—June had even gone as a student to her college's psychological counselling centre, where she was told she had emotional problems requiring therapy. Several months of treatment yielded no result and only succeeded in upsetting her more.

June unsuccessfully tried to build up a whole system of compensating methodology. She was very compulsive, wrote things down a lot and took other action that enabled her to function at what appeared to be a near normal level, albeit with tremendous internal tension, which often made members of her family nervous to be with her when she was working.

Today the change is startling. What struck me first is that June has developed a sense of humour. This is a very unexpected development. Her level of conversation is higher, and she was able to understand the main concepts of Tolstoy's theory of history and point out its shortcomings.

She takes things less at face value. She has started recognizing that people she knows have limitations. She asserts herself much more appropriately and defends her own point of view. She doesn't just follow my direction but uses her own judgment, which is now quite good. She does things less slowly than before. She handles situations in general much better. She agonizes and worries less over things and feels much better about herself.

June recently had a short telephone conversation with our daughter. As soon as she hung up, our daughter remarked that it was eerie speaking to her mother, who seemed to have changed. She found her calm and collected, never her normal characteristics.

This composure now seems to permeate everything she does. When an unexpected situation comes up, she seems to step back and analyze and act rather than getting nervous. She is in the midst of the holiday period, always a time of great tension in our house. And she is not afraid to take a day off to visit her daughter, as she did today.

Many things are just being done more simply and with fewer problems, so that we don't even notice them. Her level of confidence is higher, based on her functioning, which created a much happier outlook on life. Our life together, which was always happy, unexpectedly improved. We find ourselves spending more time together, and enjoying each other's company more.

Robert Winters

Over the years I have asked adult students about their learning disabilities and the impact these have on their lives. What I came to understand from those discussions is that a learning disability, and especially the symbolic thinking deficit, is an insult to adult autonomy. These individuals do not feel fully mature.

Ten years after completing the program, a woman wrote me an eloquent letter describing the autonomy she now experiences:

Looking back, I can only describe my life to that date as existence, struggling along trying to survive in what was for me a continuously confusing world. With graduation from high school looming, I spent a great deal of energy trying not to think about the future, because I really hadn't a clue what I wanted to do with the next sixty or seventy years of my life. Such was my desperation, I had narrowed my options down to two: find someone to take care of me, or live outside the law.

The most important change is that I now have an effective ability to learn: to apply ideas to my life, and solve problems—really solve them, whereas before they bewildered me.

I have the capacity to choose those ideas that are important to me, to discard useless information, to set my own standards and to have confidence that what I absorb is reliable and will not disappear into a mist somehow.

I have a satisfying career, which I chose because I can sum up my assets and decide which ones I wish to highlight. In other words, I have control—over my life, over what is done to me, and over where I will be in the future.

I used to feel I was living in a hurricane being buffeted by the strong winds, pulled hither and yon, and now I can stand calmly grounded in the centre of any storm and figure my way out.

Other Arrowsmith students have described a similar experience: how in the past, emotion-laden events created whirlwinds that buffeted them terribly and, by contrast, once the brain had cleared, how calmly they were able to face similar events and plan a response.

The prefrontal cortex, in both the left and right hemispheres, is central to judgment, planning, problem solving, self-direction, and self-regulation, all functions critical for full adult autonomy. And as this area improves, what we witness, as this woman describes, are individuals beginning to plan and direct their own lives effectively.

CHAPTER TEN

WORDS FAIL

Islands in the stream, cut off from the rest. This was the lot of Tanya Day, Nicolas Dalton, and Adriano Genesi, all of whom once struggled to have their speech understood.

I remember when Tanya first came in for testing. Owing to a severe predicative speech deficit, she had limited speech that I could not comprehend. Nicolas Dalton was almost totally isolated by the same deficit, one that meant in his case that no one but his family could follow him. He could not speak or write a coherent sentence; instead he made lists. Adriano Genesi was so challenged in his attempts to express himself that pantomime was virtually his only recourse.

The complexity of language dictates that numerous areas in the left hemisphere of the brain interact in elaborate ways to convert our thoughts into ordered and meaningful sequences of words. Luria, in *The Working Brain,* suggested that this deficit may reside in the inferior postfrontal zones of the left hemisphere. Research does not point with certainty to which areas of the brain are involved in the difficulties described in this chapter. Through the stories of Tanya, Nicolas, and Adriano, however, I can describe the classic features I have seen with many students and what happens as the deficit is addressed.

In his book *Traumatic Aphasia,* Luria described an advance in the development of language over the course of human history in which words moved from having just a naming function (a means to represent an image or concept) to being used to represent whole statements or thoughts. He called this latter aspect of speech, in Russian, the *ëpredicativeí* aspect of speech (all parts of a sentence beyond the subject), which is why I came to call this deficit *predicative speech.*

The individuals I saw with the predicative speech weakness all struggled to express thought in language beyond simple naming. Luria observed that a person with this problem was not deprived of words but of the ability to order words to formulate a complete thought or sentence—that is, the ability to learn syntax. The word *syntax,* from ancient Greek, means an arrangement, an ordering together. This is what these students could not do: arrange words to voice a thought. When they strung several words together, careful analysis showed the expressed thought was disjointed and incomplete. Both speech and writing were clipped, and word order was off.

The language difficulties that Tanya, Nicolas, and Adriano had would traditionally be identified as an expressive language disorder—an impairment in the ability to use spoken language to express thought at a level expected for the individual's age. Students with this difficulty may have any or all of these symptoms: a reduced amount of speech; restricted vocabulary; the use of short, simple, and often incomplete or grammatically incorrect sentences; incorrect word order; missing functional words such as prepositions; repeated use of certain expressions or phrases; and substituting general words such as *thing* or *stuff* for the more precise word.

As a result, these individuals are unable to communicate thoughts, needs, or wants at the same level or with the same complexity as their peers. The listener struggles to follow and understand their speech. Given such communication challenges, these individuals may have difficulty socially and may appear less capable than they really are. When the overall intellectual functioning of these individuals is assessed, it's important that nonverbal tests be used. Tanya, for example, had wildly varying scores on intelligence tests; when language was required, she did poorly, scoring in the 1st percentile.

On the surface, it appears that external speech is the primary difficulty, but the problem is more profound than that, as we will see with Adriano. This deficit affects not only external speech (spoken aloud) but internal speech (in our heads) as well. Internal speech is critical: it allows us to guide and regulate our behavior. Unable to rehearse inside his head before speaking, a person with this deficit is prone to saying the wrong thing, leaving him open to the accusation that he is rude and lacking tact. Given a gift, for example, someone with a predicative speech disorder might blurt out, "I already have one of those." Internal speech also allows us to repeat and recode information in order to retain it.

This neurological deficit, then, affects memory, thinking, speech and writing.

When Tanya Day was six years old, she spoke like this:

> "Where daddy went?" (She meant, *"Where did daddy go?"*)
>
> "What do you doing?" (*"What are you doing?"*)
>
> "If I look funny when I do that?" (*"Do I look funny when I do that?"*)

Her mother, Ginny Whitten-Day, a public relations consultant, had adopted Tanya as a ten-day-old infant. The little girl was quick to sit and walk and ride a bicycle; she was outgoing and friendly and would prove to be very athletic. But by the age of two, the girl had no language.

And yet there was a spark. Tanya could quickly put a puzzle together, and although she could not talk, she could play a game that required her to put pictures together chronologically to tell a story.

Tanya had many severe neurological deficits and paid a huge emotional price for them. She had tremendous performance anxiety and would become immobilized, totally rigid—not just in her thinking but in her body. Ginny reported that Tanya had a spelling test and got a 0 on it, which was strange because she had a good visual memory for words and her spelling was excellent. She didn't always know the meaning of words, but she could spell them.

"I went to the school," Ginny told me. "Tanya was upset because she got 0. And I held the paper up to the light, and I could see she'd erased everything. There were ten words on the test. She'd got down to number eight and didn't know how to spell that word. To her, that meant that she couldn't do *any* of them so she just erased them all and got 0. She could have got at least 70 percent."

Tanya had fifteen different cognitive deficits, all recorded in her file after we first saw her in the summer of 1984 when she was fourteen years old. The list makes for sobering reading because every deficit but one was rated as severe, but most significant of all was her very severe predicative speech deficit. By the age of fourteen, her language was still scrambled and idiosyncratic.

Because her mother is a writer by trade and a compulsive note taker, the documentation on Tanya is extraordinarily comprehensive. In a document Ginny called "A Mother's Observations of Her LD Daughter," she kept a diary through the 1970s and 1980s as Tanya was growing up. This is an example of Tanya's sentence structure then: "Those people. When they stole something from the shop. Those people. Those policemen. When they be bad they put them in jail." This was prompted by her seeing a policeman and attempting to establish what officers did all day.

The diary is Ginny's attempt to understand her own daughter. "Is her thought pattern as muddled as her speech?" Ginny wrote. "If so, is there some other therapy, as well as speech, that could help her collect her thoughts?"

When Tanya once said, "We didn't have any supper for a long time," she meant *we haven't eaten for a long time*. "It was actually lunchtime," Ginny notes in the chronicle. "She knows the word *eaten* and the word *lunch* and yet she couldn't recall them, so she used the most reasonable ones she could think of. I do similar things when I'm in a foreign country!"

Tanya's predicative speech problem interfered with her ability to retrieve words whose meaning she knew and with her ability to learn vocabulary. One way we learn vocabulary is through the use of words in context within a sentence. Given that the predicative speech deficit significantly impairs the ability to use and understand sentences, Tanya was limited in her ability to learn vocabulary this way. Luria had observed that someone with this deficit cannot repeat a sentence just heard (so hearing how others use language was of no benefit to Tanya) or generate a sentence on her own.

Tanya Day did live in a foreign country by virtue of her many neurological deficits, predicative speech in particular: she literally didn't speak the language. In the evenings, she would be exhausted by her titanic struggles to understand and be understood.

By kindergarten, Tanya spoke individual words but never sentences. "The only thing that could get her talking was if someone mispronounced her name and said 'Tan-ya' instead of 'Tahn-ya,'" Whitten-Day recalls. Many times her teacher would do it just to elicit a response in hopes that with practice, Tanya would get more comfortable speaking.

When Tanya was seven, a private psychologist diagnosed her troubles as "maturational lag" and told Ginny not to worry—that the passage of time would resolve everything. Dissatisfied with that advice, Ginny had her daughter tested repeatedly. The tests, she found, were consistent only in their inconsistency. "We were told," says Ginny, "she was an overachieving retardate and an underachieving genius and everything else in between."

One time when Tanya was still quite young, the family's pet hamster escaped his cage and sent everyone into a tizzy. Tanya calmly retrieved an empty margarine container and placed it over the hamster. "Someone who's not intelligent," Ginny remembers thinking to herself, "doesn't do that."

However, when it came to language, simple was manageable, and complex was not. This was why, as Ginny put it, "Tanya spoke in nouns." She could say *table* but not *eat at the table in the corner.*

Tanya came into our school with a great deal of anxiety. She had experienced so much failure, and as she would master one level of an exercise, she would express fears about going on to the next level. "Dear Miss Arrowsmith-Young," she wrote in a letter to me. "I really think I should stay on predicative speech level 8. I just really think that this other is going to be too hard for me. This is a better level." But we'd work with her, and she'd move on.

What Tanya had in her favor were extraordinary determination and an intelligence that few but her mother believed she possessed.

To address her predicative speech disorder, Tanya listened to simple but correct speech, which she had to repeat. Simple became complex, the pauses shorter. Listen, repeat. Listen, repeat. Tanya found this hard, especially as the challenges mounted. But then her speech began to change.

Six weeks into her first year at Arrowsmith, Tanya volubly declared her dislike for the *Toronto Sun.* The tabloid newspaper then featured a scantily clad young woman as part of its logo, and Tanya, an ardent feminist at fourteen years of age, strongly disapproved. "I'd like to go down to that newspaper office and ask the manager, 'Why are you putting those naked—well, bikini-wearing—women on our newspaper?'" she told her mother. Perhaps, her mother suggested, the woman was there because men like it. Tanya promptly retorted, "Well, what do women like?"

One day she was watching the news on television and someone was being interviewed behind a sea of microphones. Tanya started to laugh and said, "What would that guy do if those press people pushed the microphone too close and it went down his throat?" She imitated a coughing sound, then uttered what she imagined he would say: "My God, you've given me a heart attack!"

Tanya had never spoken in such complete sentences before. We were witnessing dramatic change in her language—from monosyllabic to the beginnings of normal speech.

Ginny wrote in her diary around this time: "Tanya had a great story today about why she'd spent her lunch money on cookies—very cohesive, perfect sequencing, very expressive face. A few weeks ago she would have just answered with a shrug and said, 'I don't know.'"

Four months later, Ginny left Tanya written instructions on how to make turkey dinner. Everything was ready on time, and Tanya had washed the dishes as she went along. Her newfound ability to understand language meant she could follow procedures and start to perform age-appropriate tasks.

A few weeks after that, Ginny recorded this entry: "Tanya discovered her brother and sister would be sleeping over at a friend's house. She said, with a huge smile and a thump on the back for her father, 'Good work, Dad. We're going to have a peaceful evening.'"

Even her younger brother, notoriously slow to compliment his older sister, was commenting favorably on Tanya's speech. Less than two weeks later, this diary entry from Ginny: "I have noticed that I have stopped making my sentences simpler for her. In fact, the other day when I started to repeat some instructions in a simpler manner, she said, 'Mummmm!'"

Here was clear evidence that Tanya's profound predicative speech deficit was starting to be addressed. As her cognitive capacity improved, her brain could now make sense of how words needed to be strung together to express a thought, and she could listen to, follow, and learn from language being used by people around her. She began to have a richer vocabulary that she could use properly in expressing her thoughts. She would have language at last.

Tanya was even able to correct her mother at times. Ginny says, "I couldn't think of the words *revolving door,* and I said *twirling door.* Tanya said, 'You mean the revolving doors.'" Ginny also recalled that Tanya's new language skills were not an entirely unmixed blessing: "She would talk Duran Duran [a British new wave band popular in the 1980s] until I was ready to scream."

Though she was a high diver from the age of eight, and early on had used play and sport to connect with friends, Tanya had become increasingly isolated socially by her language challenges. That isolation was ending. A former teacher of hers saw her in March 1985 and remarked that Tanya no longer muttered into her chest, that she now stood taller and made eye contact.

Tanya could now communicate socially. "We had a birthday party with a lot of adults and she sat and talked to people throughout the whole thing," Ginny wrote at the time. "Before, she would have retreated to her room. When she tried to communicate previously, she could not make herself understood, so she became frustrated and withdrew. She was shy with adults. And a neighbor told me until lately he hadn't realized that Tanya had dimples. Now she knows how to smile."

"I can see an enormous difference," wrote Ginny in a summary recorded in 1985, "in her understanding. If there is an area that she isn't sure about, she can ask now. Her face is definitely more relaxed. The tension from trying to understand and make herself understood seems to have gone. She is following everything and making an acceptable crack at speaking back."

Tanya reentered regular school as a ninth-grader after two years of intensive cognitive work and confidently tackled a Chekhov short story for her first literature assignment.

In May 1988, Tanya's intelligence was tested once more, and on a language-based test she was found to be at the 37th percentile—within the average range of intelligence. On the Raven's Progressive Matrices test, a nonverbal test of intelligence, she was at the 60th percentile.

Tanya's intelligence had been there all along. It was just blocked. At her old school, she looked as if she belonged in the bottom 1 percentile: she couldn't talk, and she couldn't learn language. She might get a simple question, but the minute the question had too many words, she was lost. There were times she had the answer; the question was what confounded her.

For the past ten years, Tanya has worked as an office assistant for an engineering firm. Were you to meet her for the first time, Ginny says, you would have no inkling of her past struggles, especially if the discussion was about hockey, rugby, or the British soap opera *Coronation Street* (three of her passions). Some of the deficits are still there, since we were unable to address everything in her time at Arrowsmith, but those that remain are now in the moderate to mild range.

Tanya lives on her own, manages a budget, socializes, and enjoys books and sporting events in her spare time. She and her mother discuss what they've both read in the newspaper. Tanya often says how truly happy she is to have her job, one she enjoys. Part of her job is mail delivery within the firm, requiring a cart and the building's biggest office, a source of pride for her. She likes it, too, that office mail often goes awry when she's not there.

What is certain is this: had her brain deficits not been addressed to the extent they were, Tanya would not be independent today. She would be living with her mother and not working, or living in an institution at taxpayers' expense, or living on the street with untold consequences.

To witness Tanya's transformation, to see her emerging as a person, was unforgettable.

All parents want smooth sailing for their child. They want the child to be loved, to be accepted, to be "normal," and the designation of normal, or not, is often applied within minutes of birth.

Nicolas Dalton came into this world with poor Apgar scores. The Apgar score refers to the (A)ppearance, (P)ulse, (G)rimace, (A)ctivity, and (R)espiration of the newborn. Is the infant's skin color blue (bad) or pink (good)? Is there no response to stimulation (worrisome) or a vigorous one (ideal)? A low rating could presage serious neurological problems.

Nicolas was weak, needed resuscitation, and was not responding normally during his first few minutes outside his mother's womb. He was allergic to his mother's milk, and the struggle to get nutrients into his body would continue for three years. Early on, Nicolas was given a diagnosis that his mother, Teresa Dalton, prefers not to discuss. She will only say that Nicolas was badly misdiagnosed.

At the age of three, the boy began speech and occupational therapy, and he required an educational assistant when he started school. He had no language save for one or two words strung together. Nicolas could not communicate.

He had been held back a year in school along the way. His long list of neurological deficits meant that reading, writing, speaking, understanding, and looking someone in the eye were all challenges for him. Because he couldn't follow any discussion, Nicolas was always off topic in class— so much so that one time when he did respond appropriately, his classmates all applauded. He played sports and had what his mother called "sports buddies," but there was no conversation between him and them, and therefore he had no real friends, which troubled him more and more.

Nicolas was distraught, but so were his parents. Would he ever marry? Not likely. Hold down a job? "I don't know if he could have been a dishwasher," said Teresa. "He couldn't stick with anything. He couldn't do step one, step two, step three." Without language as a tool, he could not guide his actions.

In January 2006, at the age of eleven, Nicolas was tested, and he commenced a program of cognitive exercises that spring. He was inordinately enthusiastic and initially thought he could get through everything in a month. A burden of nine deficits would not be shed so easily, but it would be shed. Just as Tanya had done, Nicolas started to work on his predicative speech disorder. He donned headphones, listened assiduously, and repeated aloud what he heard. This exhausted him, as if he'd been on a treadmill. And as with Tanya, his brain responded as more and more was asked of it.

"Early on," Teresa told me, "we began to notice little changes with his ability to understand, his ability to communicate. And when he got to the point where he could really talk, we learned that he was a very deep thinker."

When Nicolas finally began to speak, one of the things he said was this: being in the grip of this particular learning disability, he noted, was like staring into a big pot of soup full of words. He was trying to pick out those words to put a sentence together in speech, but the soup would not stop swirling and the words were all mixed together. Nicholas had come up with a very good way to describe the predicative speech deficit.

His reintegration into school after Arrowsmith presented some challenges. Not only did he have to acquire the language that his peers had spent all their lives amassing, but he had to make up ground academically and socially.

On reentry into a full academic curriculum, Nicolas was in a transition period when his brain capacity was in place but not the skill set or the academic content or the language that he had been unable to learn while the deficit was in place. Once someone's brain capacity changes, that person learns at a faster rate, but there's much missed ground to cover.

We often suggest to people in Nicolas's shoes that they engage a tutor for the first year to teach them any missed content. In addition, on completion of the program, students need to gain experience using their newly strengthened cognitive areas, and this can take up to one or two years— just as someone who has had a surgical procedure to fix a lifelong limp will need time to learn to walk naturally.

Teresa put it well: "It was like Nicolas was going from outer space to the land of humans."

Remarkably, or perhaps not so remarkably, Nicolas began to soar. He went from being a C student at the beginning of eighth grade to the top of the class. After graduation, Teresa and her husband put Nicolas in the International Baccalaureate program, which meant an accelerated and demanding academic load. His average grade at the end of ninth grade was 90.8. Nicolas's dream was to go to a certain academically rigorous prep school, and that's where he is now. His second-term average was 85.

Like many other parents who land in the Arrowsmith stream, Teresa never lost faith in her child. As for those diagnostic labels applied in the first few years of his life, she said, "I knew that it wasn't correct because I would see little glimmers of hope." She agreed that Nicolas had been on a great journey, a long and arduous one, but one that had "molded and shaped him into one of the most amazing individuals."

He has told his mother that were it not for Arrowsmith, he'd be dead now. "He meant," said Teresa, "that he would die in sorrow. He was so distraught. Our life has totally changed. When parents come to me and ask me about Arrowsmith, I always tell them the same thing: 'If you don't pay now, you're going to pay later—and dearly.'"

I was very moved by Nicolas's story. He was trapped, had no language, could not communicate with the world. And then he came out, almost as if he had been in a coma.

"It's like an awakening," Teresa said.

Individuals with a predicative speech dysfunction struggle to express external speech but they also suffer from a paucity of inner speech—the silent dialogue we carry on inside our heads. A child speaking aloud uses this self-directed talk to determine whether to carry out an action and, if he decides to proceed, to talk through an action. In the course of normal development, at age seven or eight, such external talk becomes internalized.

Properly functioning inner speech allows us to mentally rehearse what we want to say before we say it and to review courses of action beforehand. It thus allows us to guide, direct, and control our behavior.

Lev Vygotsky, a colleague and mentor of Aleksandr Luria, made a special study of internalized speech, which the latter continued to investigate after the former's untimely death at the age of thirty-seven. Inner speech, Vygotsky said, "is speech for oneself: external speech is for others." He viewed inner speech as a kind of shorthand—one that is highly abbreviated and devoid of subjects. "It is as much a law of inner speech to omit subjects," he wrote, "as it is a law of written [and spoken] speech to contain both subjects and predicates."

Individuals with the predicative speech problem have particular difficulty with predicative elements in both external and inner speech.

I first met Adriano in 1978. At the age of eleven, he had almost no external speech and, as far as I could determine, next to no internal speech.

He could not speak in sentences and could utter only disjointed words accompanied by gesturing. His mode of communication was to say one word, then act something out, then say another word, followed by more acting out. Because he acted everything out, officials at Adriano's elementary school had told his parents that he would never learn to control his behavior and could not be educated. When his father brought him to me, he said, "I know my son is brighter than that."

Guided by Vygotsky and Luria, I hypothesized that Adriano's predicative speech dysfunction made it nearly impossible for him to communicate or to control his actions. This made sense, as Vygotsky had argued that humans control their behavior with inner speech. For Adriano, I first created the predicative speech exercise—one that involved listening to and internalizing speech. We started with very simple exercises and progressed slowly to more difficult ones.

I remember vividly the day I walked into his classroom and he looked at me and said out loud: "I know, stop talking and go back to work." He could now use external speech to direct his behavior.

"Adriano," I told him, "if you keep doing that exercise, the day will come when you will be able to say that inside your head and I will never have to ask you to go back to work again."

Three months later, that day came. Adriano had reached a level of competence in his predicative speech function that meant he stopped being a discipline problem. With further cognitive stimulation, his external speech became internalized.

Adriano, Nicolas, and Tanya, once cut off by their predicative speech deficits, have joined the conversation.

CHAPTER ELEVEN

LEAP BEFORE YOU LOOK

In the thirty-four years that I have been dealing with learning disabilities and understanding how they play out, certain moments are crystallized in my mind. This is one.

A twelve-year-old girl named Vanessa is flipping through a copy of *National Geographic* magazine when she loudly exclaims, "Oh, gross!" The picture that has disgusted her is of sheep farmers in New Zealand. They are eating lunch while leaning against burlap bags full of sheared sheep wool, which has spilled onto the floor.

"This is terrible," says the girl. "These people are eating lunch sitting on dead mice!"

The year is 1983, and Vanessa is in an Arrowsmith class. She's there to address a severe artifactual thinking weakness. This deficit has several impacts, but one of them is to impair the ability to read facial expressions and body language. Those with this deficit can neither read others nor adjust their own behavior accordingly. If the proverb "Look before you leap" urges thinking first and acting second, this deficit triggers the opposite. They are likely to say or do something on the slightest of information, which means their interpretation is often wrong. (It can be comically wrong to some observers, but there's nothing funny about this deficit for those who have it or care for someone who does.) This was typical of Vanessa's approach to the world: she would take a cursory look at what was going on around her and come up with an interpretation based on incomplete information.

So how does a tendency to misinterpret what is happening in a photograph translate into everyday life? Near the end of one particular day, Vanessa saw her teacher sitting at a desk with students working around her. The teacher was tired and sat slouched in her chair.

Vanessa was too far away to hear the teacher speaking, but what struck Vanessa was the teacher's posture. That night she went home and told her parents that her teacher was sleeping in class. This led to a call from the mother, who knew her daughter's predilection and had her doubts about the story. I had to explain that, no, the teacher had not been asleep and Vanessa had not looked at all of the information before interpreting the situation.

For the same reason, parents in her neighborhood had forbidden their children to play with Vanessa. She had earned a reputation as a teller of tall tales.

If you have a strength in this part of the brain, it's difficult to imagine having a weakness here. Try this. Imagine going to Japan, where you know neither the language nor the body language. You don't have a guidebook defining the gestures, only rules of behavior you have learned from your own culture. Using those rules, you approach a Japanese person, look her directly in the eye, and put out your hand to shake hers, maybe even touch her shoulder in a gesture of warmth and camaraderie.

These are all faux pas in the Japanese culture. In Japan, it is considered polite to keep one's limbs close to one's body, so broad gestures and public touching are interpreted as rude and aggressive. Holding the gaze of another person is considered offensive; averting one's gaze signifies respect. More personal space is expected between people engaged in a conversation than in North America. Bowing on meeting is more accepted than shaking hands, and how long the bow is held and the depth of the bow all have different meanings. As you, the outsider, observe, you probably miss the foreign culture's more subtle cues.

The gestures, facial expressions, voice tone, postures, and eye movements that convey meaning can be vastly different from one culture to another. Each culture has its own vocabulary of movement that someone from another culture is blind to.

This is the daily experience of those with an artifactual thinking deficit: their neurological weakness induces a kind of social blindness. Being brought up in a familiar culture is no help to them because they cannot learn these cues from experience. They constantly make errors when interpreting nonverbal communication as if they were indeed in a foreign country and trying to decipher an unfamiliar language.

A hallmark of the artifactual thinking deficit is socially inappropriate behavior. Imagine you're at a conference and you are not fully engaged but are committed to being there. You may have an impulse—to make a phone call, leave the room, start munching on that little bag of potato chips in your briefcase. These are normal impulses, and most of us are able to control them. We don't snack loudly because we are fully aware that this would be seen as an affront to the person speaking at the podium and an annoyance to others in the audience—including, perhaps, your employer. Someone with a severe artifactual thinking deficit lacks such impulse control. Out comes the cell phone or the crinkly package of chips. This person is unable to consider the impulse, decide whether to act on it, or weigh the consequences before acting.

I chose the term *artifactual thinking* to describe this deficit, which is rooted in the nonverbal realm of the right hemisphere of the brain and, more specifically, the right prefrontal cortex. The left hemisphere is the world of symbols and language, and I labeled a weakness in the left prefrontal cortex a *symbolic thinking deficit* (described in Chapter 9). The two kinds of thinking, one verbal and one nonverbal, are parallel processes. With the term *artifactual* I wanted to encompass the entire world of things, a world beyond naming or numbering. In hindsight, "nonverbal thinking" would have been an apt description, one that encompassed the broad nonverbal realm beyond artifacts.

We live in a world of language, so we inevitably find ourselves in social situations trying to find words for our perceptions and feelings. As such, I would say that the left prefrontal cortex plays an important role in all the right prefrontal cortex tasks that are described in this chapter. Today, researchers are investigating the functions of the many subregions within the prefrontal cortices in both hemispheres and their complex functions and interactions. But when I started this work in 1978, the right hemisphere of the brain was still something of a mystery. In his book *The Working Brain*, Luria devoted a mere twelve pages to it, saying that research related to right-hemisphere functions had been neglected.

Long thought to be "lacking generally in higher cognitive function," the brain's right hemisphere is involved in many complex tasks: spatial perception, nonverbal communication, object and facial recognition, self-awareness, consciousness, empathy, perception of emotion, humor, moral judgment, music appreciation, and the intonational aspects of speech. It appears passive because it is relatively mute. It can think, process, and feel, but it cannot communicate through language.

An early and prescient study pointed to the importance of the right prefrontal cortex in initiative, planning, and evaluating courses of action. Wilder Penfield, an eminent Canadian neurosurgeon, described what happened to his sister, Ruth, who in 1928 had a tumor as well as a substantial part of her right frontal lobe removed. Penfield wrote: "Ruth was conscious of not being alert mentally: 'Each time I feel encouraged,' she said, 'I do a series of very stupid things.'"

Penfield observed no change in his sister's memory, insight, and conversation, but he did notice that she did not discipline her children and that managing the household was proving difficult. Loss of the right frontal lobe had also robbed Ruth of her capacity to plan: "Fifteen months after the operation," he wrote, "she had planned to get a simple supper for one guest (WP) and four members of her own family. . . . When the appointed hour arrived the food was all there, one or two things on the stove, but the salad was not ready, the meat had not been started and she was distressed and confused by her long continued effort alone. . . . The element which made such administration almost impossible was the loss of power of initiative. . . . She had become incapable of discerning for herself possible courses of action so that she might choose. If others presented the possibilities, she made up her mind quite easily." What Penfield is describing here is the executive role this area plays in planning.

In addition to the executive planning function, the right prefrontal region plays a crucial role in establishing the relationship between the self and the world.

Julian Keenan, director of the Cognitive Neuroimaging Laboratory at Montclair State University in New Jersey, studies neural activity related to self-awareness. He refers to the role of the right prefrontal cortex in self-awareness as "cognitive Goldilocks." This part of the brain allows us to project ourselves into various scenarios before acting and to evaluate beforehand the pros and cons of different actions in different situations, doing mentally what Goldilocks did in the fairy tale: exploring various scenarios to find the one that was just right for her.

Kai Vogeley, a German researcher, has used neuroimaging techniques to demonstrate that the right prefrontal cortex is critical for tasks that involve thinking about one's own thinking (self-awareness) and someone else's thinking (theory of mind). The latter allows us to take into account someone else's perspective in order to attribute opinions, feelings, attitudes, and intentions to others. Such awareness is an essential component of successful social interaction, allowing us to make predictions about others' behavior. Such knowledge can help us decide which of these perspectives (self or other) we should take in a particular situation (depending on the desired outcome). Someone with a problem here will either miss or misinterpret others' intentions, misunderstand how his or her own behavior affects others, and have a difficult time with social reciprocity.

As a result, this cognitive problem may seem like an emotional one and is often treated with therapy. Current psychological models tend to assume that those who have trouble reading or regulating their own emotions or reading those of others have these difficulties because of psychological conflicts. What I have seen, however, is that many people who seem to lack emotional intelligence actually have weakened functioning in the right prefrontal cortex, and they can improve only by working on this cognitive area.

Neurological deficits can play havoc with parent-child dynamics, some deficits more than others. Perhaps the most ruinous in this regard is the artifactual thinking weakness—the social deficit, I sometimes call it.

Consider the case of Nathaniel Freeman. He's in tenth grade now and thriving in a competitive academic Toronto school thanks to several years of dedicated cognitive work. But for the first ten years of his life he paid a terrible price for his deficit.

Six years ago, the psychometrist at Arrowsmith School asked Nathaniel to complete some sentences. "I want to know . . ." the first sentence began, to which he added ". . . how smart I am." The aim of the exercise was to acquire a sense of how he feels about himself, and what the simple test revealed was the measure of his misery. Nathaniel described how "people hurt me for no reason at all" and that he feels "sad when everyone criticizes me." He wrote that he hates school, does nothing, that his so-called friends hurt and bother him, and that he has failed countless tests. One of his teachers observed how painful it was to watch Nathaniel walking down the school corridor: "It was like the parting of the Red Sea."

He had one friend, a child who had the same artifactual thinking deficit that Nathaniel did, but even more pronounced. Misery literally had company.

With this deficit, there is no "off" button. Nathaniel's questions never stopped, and I remember all too well how exhausting it was being around him. Unable to read facial and body cues and focus on anything beyond their own immediate needs, people with a significant artifactual thinking deficit have little or no capacity to make and keep friends, and they soon find themselves outcasts. Neither Nathaniel nor his parents had a clue as to what was causing him to be the way he was (other than the thought that social awkwardness ran in the family). His parents simply assumed that the son they dearly loved was also exasperating in the extreme, and nothing could be done about it.

Nathaniel's younger sister, Abbie, had brain deficits of her own, but she possessed a great gift for the social graces. Nathaniel complained that she got preferential treatment, and his parents now concede the truth of that charge. While he railed and demanded, and usually got nowhere, she asked sweetly and chose her moments carefully, and she often got what she wanted. Brother and sister at home would commit the same minor infraction: he got punished, she got a smile.

Hannah Freeman says Nathaniel frustrated her all the time. As a result, she says, "He absolutely got horrible, horrible feedback from his teachers, his friends, his father, me. I feel incredibly guilty about it. I remember when he was tested, and we reviewed the report and then you, Barbara, were explaining the artifactual thinking deficit to me and I said, 'Oh my God, I've been yelling at this child for years, and it's not his fault.'"

Nathaniel could tell time, but if Hannah was out and told her son she'd be home in an hour he'd call her every five minutes during that hour. He focused entirely on the immediacy of his own needs and was unable to consider the impact he was having on his mother. By the time Hannah got home, she would be so frustrated that whatever she'd promised to give Nathaniel was no longer on the table. There was tension between mother and son and between father and son. Both parents were hard on Nathaniel.

"I'm a mother," Hannah explained to me. "I shouldn't be impacted by how my son asks for something. I should give whatever it is to both my children equally, no matter what. But I didn't."

Much has changed for Nathaniel. Two years of the cognitive exercise to address the artifactual thinking deficit have induced a remarkable change in Nathaniel himself and in the family dynamic. He now has numerous friends, a lively Facebook page, and, thanks to work on his other deficits, he's doing extremely well in school.

"A happy child," is how one teacher describes him. Children with this deficit are prime targets for bullies, and it has been a relief for both Nathaniel and his parents to be rid of that worry at school.

"A sweet kid with a big soul and a good heart" is how Hannah describes her son. He was always a gentle and thoughtful spirit with a philosophical bent, but his inability to filter his thoughts and manage his emotions meant that his true nature was visible only on rare occasions. Hannah had never stopped loving her son, of course, but now she actually understands him. Much of the anger and sense of injustice that defined Nathaniel have dissipated, and his parents could not be more pleased.

"Whenever he didn't get his way," Hannah told me, "we tried explaining to him why he had to wait his turn or share, but it was almost impossible to reason with him or say no because he couldn't wait to hear an explanation. He would start sobbing, and it would take a long time to calm him down to have a conversation so he could accept the outcome of a decision. He could go from calm to a meltdown in ten seconds or less."

Now when Nathaniel wants something, he's no longer like a hammer banging loudly on a nail. "After two years of the exercise," says Hannah, "the contrast was so great that at first I was pleasantly surprised each and every time he graciously accepted my decision. He is now a pleasure to talk to and contributes ideas when decisions need to be made. He is the first one in the house to compromise when disagreements arise and he can always be counted on to understand every single person's point of view."

The artifactual thinking deficit is an especially difficult one to have because it can make you appear self-absorbed, selfish, and insensitive.

Hannah Freeman came to understand that sometimes unwanted social behavior is actually a learning disability, just like the inability to read. Armed with that insight, she was able to feel new depths of compassion for her son, who, with the aid of cognitive treatment, gradually addressed the weakness in his brain and stopped the behavior that had so alienated all those around him.

Nathaniel can now anticipate what others might be thinking and feeling. His mother once spotted him lying on his bed at midday, and he knew by her look that she was worried about him. Before she could utter a word, he said to her, "I'm not sad. Just bored."

Nathaniel can now tell by the tone of her voice when she's on the phone whether it's a personal or business call and whether it's appropriate to interrupt or not.

If he sees her hauling groceries from the car, he will help her. "He was never conscious of those things before," his mother says. "I could have been carrying fifteen things and he wouldn't have noticed."

These stories all illustrate difficulties with reading nonverbal cues, suppressing impulses, understanding others' viewpoints—and the resulting changes when the deficit is addressed. Not commonly understood is that the artifactual thinking area also plays a very important role in thinking and planning in the nonverbal world.

As Nathaniel's capacity to think nonverbally improved, he was able to problem-solve without help from his parents. On one occasion he was babysitting his younger sister but wanted to go to a party. The old Nathaniel would have phoned his parents in a panic. The new Nathaniel arranged for his younger sister's piano teacher to come early. Problem solved. Later he wanted to buy a computer, and to fund the purchase he got part-time jobs, anticipating challenging interview questions beforehand and preparing responses.

This boy who used to chew his shirt collar and cuffs from anxiety doesn't do that any more. Teachers at his old school thought he had attention deficit disorder and allowed him to walk around the classroom every few minutes; teachers at his new school treat him like every other capable student. They have no idea Nathaniel ever had a learning disability—and why would they, with his 92 average?

Nathaniel's ambition is to become an engineer, although lawyer and psychologist are other possibilities now that he can plan effectively. In a composition he wrote in fall 2010, he quoted former major league pitcher and manager Tommy Lasorda: "The difference between the impossible and the possible lies in a person's determination." And former hockey star Wayne Gretzky: "You'll always miss one hundred percent of the shots you don't take."

There are deficits, and then there are deficits. What I would come to understand during the early days of treating learning disabilities is that weaknesses in the right and left frontal lobes (artifactual thinking and symbolic thinking) are the critical ones. These deficits trump others elsewhere in the brain.

Artifactual thinking and symbolic thinking work together and complement each other, enabling us to think, plan, problem-solve, and generate strategies. A strength in these two areas of the brain better equips one to compensate for other neurological deficits, and a problem in either area serves to magnify the impact of all other learning problems. These two areas are important for self-reflection, and the less self-aware you are, the less able you are to understand that you have problems requiring solutions.

Nathaniel Freeman is a good example of this. We did not have time in his training to address his kinesthetic deficit, so his clumsiness is still an issue. Before we addressed his frontal lobe deficit, not only would he be unaware at the table of where his elbow was relative to his glass, he'd be unaware of the mess around him. When the mess was pointed out to him, he couldn't understand that there was anything he could do differently. Now that his nonverbal awareness has improved, he sees the effects of his kinesthetic deficit and devises compensatory strategies: Nathaniel leans in over the table when eating and always positions his glass in such a way that it is less likely to be knocked over.

Elena Andreou was in the habit of warning teachers before school started that her son, Dimitri, sometimes blurted out inappropriate things. When the teacher, for example, promised the class that he would take the best-behaved pupils to the local pizzeria at the end of that week, Dimitri loudly announced that he hated the place. He later very much regretted the outburst. That was the pattern: outburst, regret, outburst, regret. Dimitri was like a car without brakes.

Dimitri had a more severe artifactual thinking deficit than some but less severe than others. Someone with average functioning in this area would have filtered the impulsive thought that popped into Dimitri's brain and not stated it out loud to the class, realizing beforehand the negative impact it would have. At the milder end of the spectrum, individuals may not anticipate reactions by the teacher or the class but would have noted the mistake by reading the teacher's face and immediately apologized. Dimitri could neither anticipate the response nor read the teacher's reaction. His mother needed to explain his mistake to him. Fortunately, Dimitri's deficit was not as severe as some other former students of mine who would not have understood their mistake even once explained to them.

For someone with an artifactual thinking deficit, a classroom can be a terrifying place. Dimitri knew that, and so did Claire Shapiro, who told me this story. She was sitting in her high school accounting class, this day being supervised by a substitute teacher. She had no aptitude for numbers, and her regular teacher had given her permission to work on her English homework while the others learned how to balance books.

"What are you doing?" the teacher asked Claire.

"My English," said Claire, who never thought to explain why.

The teacher read sauciness into that remark, and said, "You can either stop doing your English now, or you can leave this class."

Claire took this directive literally. She thought she had to make a choice, so she left the class. The teacher pursued her down the hall and said, "Where are you going?"

Claire said, "I thought you told me I could leave the class."

Someone with an artifactual thinking deficit can be intelligent but do poorly because he or she does not pick up on cues from the teacher or try to please that teacher or attempt to understand that teacher's expectations. Likewise, at work, such a person can be a good computer programmer but cannot read the boss's cues and therefore doesn't know when it's inappropriate to ask questions. While programmers may be able to keep a job with this deficit, a negotiator or a salesperson would not. People in these jobs need to read reactions and modify their own actions accordingly. Similarly, a doctor with this problem will have a poor bedside manner and lose patients.

People with an artifactual thinking deficit lack discretion and make bad confidants. Unable to size people up, they are poor managers. At a dinner table, they will talk your head off. They are quite unaware of their own appearance—the mismatched socks, the unbrushed hair. One mother of a child with this deficit described being in the middle of yet another loud and embarrassing intrusion by her daughter in a public place and giving her a little "let's be quiet" pinch. She isn't proud of it, and freely admits it backfired when her daughter shouted, "You pinched me!"

Looks can be deceiving. What seems like irresponsible behavior may well be an artifactual thinking deficit. And sometimes that behavior appears comical, sometimes inappropriate, and sometimes just plain reckless.

An Arrowsmith student had been banned from a certain grocery store because at unmanned snack stations (where new cheeses or crackers, say, are offered as samples) he was in the habit of nibbling—and not stopping until everything was gone. I remember a college student who did not think to take the shoe for his right foot to the clinic when he got his cast removed—so he had to hop home with one foot bare. I know a woman who put a cup of boiled water on a table within reach of her child: she could not anticipate that the baby might reach for the cup and scald herself, which is just what happened.

On occasion I have met with parents who are baffled at their son's or daughter's inability after graduating from high school to apply for college within deadlines or to look for a job—all the things we expect of an independent, self-directed young adult. Bereft of initiative, these individuals cannot do all the things necessary to be independent in the world. I have seen adult children of average intelligence with this deficit who have had to move back in with their parents.

The right prefrontal cortex allows us to navigate our social and emotional world. Without this capacity in place, we are rudderless. This learning disability plays out in homes and workplaces as well as the classroom. An inability to put oneself in another's shoes, suppress impulses, and consider another's feelings: all this ends up fracturing friendships and relationships.

When Jessica Graham was four years old, her parents brought her into a high school gym. She saw a rope and, before anyone could stop her, she had climbed it all the way to the ceiling. When she talked to anyone, she stood inches from them, unaware of the discomfort she was causing. Sleep came, but only after she had been allowed to burn off tremendous amounts of energy. The only way to tire out little Jessica and convince her it was bedtime was to let her chase her dog, a miniature greyhound, for about forty-five minutes. The family built a cedar rail fence around their lot—not to contain the dog but to contain their daughter. The girl's high energy, coupled with her inability to control her impulsivity, made her mother worry for her daughter's immediate safety, never mind her long-term future.

Her mother, Laura, noticed that Jessica couldn't grasp the notion of being made to sit on the stairs for a time-out. Unaware that she had done something wrong and was being disciplined, she would just pop up like a jack-in-the-box. Nor could she follow or learn basic instructions about road safety, so her older brother became her protector. Jessica's father would coach her in ice hockey, and he would tell her during practice to find a spot at the end of the line so she could learn by observing others doing a certain drill and follow their example. But told to step *over* sticks laid on the ice, she would step *on* the sticks (and not fall because her balance was exceptional). These were all telltale signs of a severe weakness in the right hemisphere's prefrontal cortex.

Jessica Graham was oblivious to the world, but not because she was deliberately disobedient or oppositional. Though this behavior may signal an emotional problem, it can also suggest a cognitive deficit.

Young children learn appropriate behavior through their interaction with their parents. When a small child does something—for example, if he is able to read his mother's body language and see that his mother's response is not favorable—he will generally change his behavior. When a child is older and a mother tells the child that she is not pleased with his actions or does not respond favorably to his request, a child with average ability in this area will learn to change his behavior next time to achieve the desired outcome.

A child with an artifactual thinking deficit, however, is not able to learn appropriate behavior or modify her own behavior from experience. This area of the brain is essential to learning which behaviors are appropriate, or not, in different social contexts. What is fine at home may not be fine at school; Jessica could not grasp the difference.

Through her experience in summer jobs and volunteering in schools, Laura Graham had developed an interest in learning disabilities, and everything she was reading in the literature only heightened her alarm about Jessica. Like many other parents of children with this neurological deficit, Laura's big fears were these: When this girl becomes an adult, how will she get on in the world? Will she be able to hold down a job? Support herself? Will she always be under our care?

At one point, Laura took her daughter to an audiologist, praying that what ailed her daughter was a hearing problem. That, she thought, was an easier fix than this—whatever "this" was. Some days, Laura wondered if Jessica had an attention deficit hyperactive disorder (ADHD).

Students with problems in either artifactual or symbolic thinking are frequently identified as having this disorder since the symptoms—lack of focus and impulsivity—are similar. Studies have shown that some children with ADHD have a reduced gray matter volume in specific parts of the brain (the prefrontal cortex in particular) and that their frontal lobes are less well connected with other regions of the brain.

At Arrowsmith, what we have found in the majority of these cases is this: as the underlying cognitive problem is addressed, the ADHD symptoms are also addressed, and individuals on medication are able to discontinue these medicines.

Jessica was five years old when she came to us for testing. Testing identified a photographic visual memory that allowed Jessica to teach herself to read at the age of four. But we also found eight different brain deficits, many of them severe, including an artifactual thinking deficit.

Jessica's thoughts, I remember, were disorganized. During testing, I would show her a picture of, say, a bear, and she'd correctly identify that animal and tell a story. I'd then move on to the next image, a different one, and ask her about it. "A bear," she would reply. Still fixated on the previous item, she could not reorient herself to the task at hand. Asked to comment on an image of a boating accident, she said they were playing basketball. This is a classic illustration of the artifactual thinking deficit: in the picture were people wearing swimsuits, which Jessica took to be shorts, and quickly (too quickly) she deduced that a basketball game was underway. Like the girl looking at the photo of the sheep farmers, Jessica hadn't looked at all the details to get the big picture, only a few, so she had leapt to the wrong conclusion.

In his book *Higher Cortical Functions in Man,* Luria recorded the eye movements of patients with right frontal lobe lesions when they were looking at a complex picture. These individuals, he found, "do not make the necessary searching movements with their eyes; they pick out just one detail (which they fix on for a very long time) and, on this basis, they guess the meaning of the picture as a whole; later their eye movements either continue to fix this same detail or they follow a chaotic course."

The eyes are fine; the problem lies with the right prefrontal cortex and its failure to direct the surveying of this particular picture and, in general, the entire nonverbal world. People with this deficit will have accidents not because they have a kinesthetic deficit or are uncoordinated but because they are inattentive and unaware. One student of mine had had multiple car accidents as an adult due to an inability to stay present and alert to traffic and pedestrians. Safe driving means taking in all the details, and she could not—a telltale sign of this deficit.

During one test, Jessica Graham was asked to look at four pictures and identify the one that represented a word or a concept. Disappointment, for example, was illustrated by a child reaching into an empty cookie jar. Jessica would look at one or two of the four items and make a choice, or just pick all of the same item number for several questions—say, answer number 3. When left to her own devices, she scored in the 19th percentile (low average). But when I gave Jessica an alternate form of the same test (she was to point to each item in sequence and then wait five seconds before making her choice), she scored at the 62nd percentile (average). That's a 43 percentile difference. Jessica's artifactual thinking problem was interfering with her ability to survey the material. Her impulsive, scattered approach was her default mode, leading to performance well below that expected given her intelligence.

I have had other students like Jessica—students with severe deficits in both the right and left prefrontal regions. These two areas of the brain drive the whole system to engage in the learning process. Unable to engage sufficiently, children with problems in these two areas cannot provide the ongoing stimulation necessary to drive intellectual functioning. One minute a bird has their attention, then a cartoon, then something else. I have observed several cases in which this random approach to learning led to IQ scores that actually decreased over time.

To help Jessica see, truly see, her world with focus and attentiveness, we assigned her the task, as we do for anyone with an artifactual thinking deficit, of examining narrative art—illustrations that tell a story or convey a parable. This process of active and purposeful exploration is guided by the prefrontal cortex in the right hemisphere and is similar to the text exercise described in Chapter 9 used to improve symbolic thinking.

In this exercise, the student strives to understand the picture's storyline by carefully considering the dynamics between the individuals portrayed. The aim is to come up with hypotheses that make the most sense while discarding ones that make the least sense before reaching a conclusion about what this piece of art is saying.

The aim of this exercise is to stimulate the prefrontal cortex so that students can eventually read nonverbal cues on their own, be aware of their social world, and plan their actions accordingly. There is no rule book for life and no set of rules comprehensive enough to govern every situation. It is critical, therefore, that individuals learn to think for themselves. As I hypothesized in 1980, these individuals not only get better at reading others' nonverbal cues but also become more self-aware. As with all the other cognitive exercises, this one targets the broad range of functions governed by this particular cognitive area.

In 2009, Jessica was tested by her school's psychologist, who could find no trace of a learning disability. The Arrowsmith Program, says her mother, "gave her life." Without this intervention, her mother is convinced, Jessica would be on medication and in a behavioral class.

That "pinball of energy," as Laura once described her daughter, is in control now and is aware when something she's done makes people angry. She has become the captain of her university basketball team and skillfully mediates team conflicts. Once a scattered little girl, she now has emotional intelligence, focus, and discipline in spades.

"She is a beautiful young woman now," her mother says. "Fully aware, socially poised, and adept."

CHAPTER TWELVE

WHEN A PICTURE DOES NOT PAINT A THOUSAND WORDS

"Good-bye, till we meet again!" she said as cheerfully as she could."

"I shouldn't know you again if we did meet," Humpty Dumpty replied in a discontented tone . . . "you're so exactly like other people."

"The face is what one goes by, generally," Alice remarked in a thoughtful tone.

"That's just what I complain of," said Humpty Dumpty. "Your face is the same as everybody has—the two eyes, so—" (marking their places in the air with his thumb) "nose in the middle, mouth under. It's always the same. Now if you had the two eyes on the same side of the nose, for instance—or the mouth at the top—that would be some help."

—LEWIS CARROLL, *Through the Looking-Glass*
(1871)

The eminent neurologist and author Oliver Sacks suffers from a neurological condition called prosopagnosia, or "face blindness," and he wrote about the darkly comic side of this problem in 2010 in a *New Yorker* piece aptly titled, "You Look Unfamiliar." The term *prosopagnosia* derives from the Greek words *prosopon* ("face") and *agnosia* ("lack of knowledge"). Depending on the degree of impairment, this quite common yet much misunderstood problem may mean you have any or all of these difficulties:

You cannot recognize the faces of family and friends out of context.

You cannot recognize, even in context, the face of a familiar person.

You cannot discriminate between unfamiliar faces (so that, shown a series of photos of faces, you are unable to say whether the faces are the same or different).

You cannot distinguish a face from an object.

You cannot—and this is the aspect of face blindness that some find the most shocking—recognize yourself in a mirror.

The condition is often, but not always, accompanied by recognition impairments involving the world of objects. We now know that there are separate cortical areas—some that preferentially recognize objects and others that preferentially recognize faces—and that individuals can have problems in recognizing either. The majority of cases we work with at Arrowsmith report problems with both and describe the visual world as flat and not visually memorable; these individuals don't perceive visual details. Sacks describes having difficulty recognizing both faces and places (including, in one unusual circumstance, his own house).

Thought to govern face recognition is a network of areas located primarily in the occipital temporal regions of the right hemisphere—the right occipital face area, the right fusiform gyrus, the right fusiform face area, and the superior temporal sulcus. When they're functioning properly, these areas of the brain process information in order to answer a series of questions:

Is this a face?

Are these the same or different faces?

Is this the face of someone I know?

Is there an identifiable expression on this face?

Possibly because facial recognition is so important for social interactions and is used intensively from birth, a large proportion of neurons in the occipital-temporal region of the brain became specialized for identifying faces. Researchers are also looking at the occipital-temporal network's involvement in object recognition. What is known is that the right lateral occipital complex in this network is active during the process of recognizing both faces and objects.

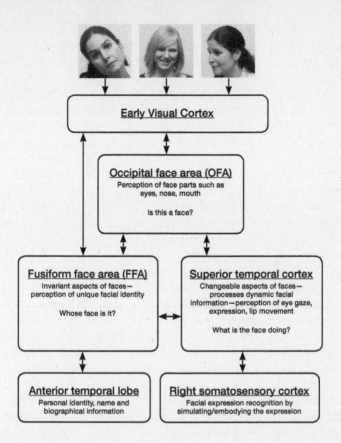

Figure 3.
A model of the face-processing network in the brain. (Adapted with permission from Pitcher et al., 2011.)

On the street, Oliver Sacks would not recognize his own assistant or his own therapist even if he had just seen them five minutes beforehand. He talks about one face-blind man who described his entire life as "nothing but apology and offense." Imagine the offense that some individuals would take at being forgotten like that; imagine the profuse apologies and attempts to explain.

In the language of Arrowsmith, face blindness is simply one form of a neurological weakness we call *object recognition*. Janice Mawhinney has an object recognition/face blindness problem and she knows the offense/apology routine all too well.

Janice came to us in 1990 with several neurological deficits, which we addressed. The one deficit we didn't have time for was object recognition. To her great distress, the objects she has all her life failed to recognize are the faces of people who should look familiar to her but don't. Nor could she recognize or remember objects, including the plants in her own garden.

Sixty-five-year-old Janice Mawhinney is a journalist who worked as a reporter at the *Toronto Star* for some thirty-eight years. She's smart and intuitive—what some might call "a seeker" (she covered municipal politics and social services but most liked writing about metaphysics). Yet her own son's face is sometimes a mystery to her.

"I was driving down a nearby street recently," she told me, "and I saw my son walking, and I was really excited because he lives in a different part of town. And I thought 'Oh, I'll surprise him. I'll pull into the next side street and say hi to him as he's walking by.' Well, when he got to the corner, it wasn't Michael at all."

Janice can look back at instances like this and laugh at some of them, but there has been an impact. "They weren't funny at the time," she told me. "They were very, very painful. Everywhere I went, all my life, I snubbed people. And the cumulative effect really was to lower my self-esteem. I keep offending people I know by not recognizing them when I see them, and that just happens in all areas of my life. And then, of course, they don't like that, and I feel that, and then I feel really guilty and badly."

She tells a story to illustrate: "I was in therapy for a while, and I related to this therapist. I loved what she was doing. At one point, I was grocery shopping, poking among the lettuces in the produce section when somebody said, 'Janice, hi!' in a really friendly, warm way. I looked around, and I had no recognition of this woman. I thought, 'Who on earth could be greeting me in such a fond way?' She could see that I had no idea. I mean, I just looked desperately blank. And she told me who she was; it was my therapist. I just felt so mortified."

Her sons have learned to help her in subtle ways. If a longtime friend of theirs—someone who has been to the house many times in the past—comes to visit, one son might say, "Mom, isn't it great that Adam made time to come and see us?" And Janice will say, "Adam, how lovely to see you again." The truth is, she wouldn't, as it were, know him from Adam.

One consequence of her condition is that Janice seldom goes to parties, which, as far as she is concerned, are hothouses of offense and apology. But a friend of hers really wants her to be there for the special parties. This friend would prep Janice and remind her who was who, but at the most recent party, the friend was occupied and there was no prepping. Janice encountered someone she knew well—someone whose son had died tragically and had been featured in one of Janice's stories in the *Star*.

"I ran into this woman," said Janice, "and I had no idea who she was. And I could see that she was hurt. Then I looked at the rest of the room, and I knew that I'd met all of these people before, and I didn't know any of them. And I was terrified. I went upstairs into the study on the second floor where it was all dark, and I sat there just hoping that the party would end and I could leave. My friend's son came up to check his e-mail, and walked into the study and turned the light on, and I was sitting there in this chair in the dark. I was just mortified, and he was stunned. My friend hasn't forced me to go a party ever since then."

During her years at the *Star*, Janice had a reputation for being standoffish, which she wasn't and isn't at all. She would enter the elevator and smile at everybody "because I knew that I might know some of them, or all of them, and I might not know any of them. And I'd rather the ones I don't know think I'm weird than the ones I do know think that I'm not recognizing them."

Once after yet another case of mistaken identity—this one involving two newspaper colleagues, both of them South Asian—a coworker was so shocked that she practically accused Janice of racism.

"Do all brown people look alike to you?" the woman asked her.

"It's way worse than that," replied Janice, who explained that all faces, regardless of color and unless she has seen them hundreds of times over the course of many years, look the same to her. The woman laughed. "I think she believed me," said Janice. "I hope she did."

Janice had compensated for her deficit as best she could, but the work was exhausting.

Take, for example, her love of gardening. Janice had written about gardening and has a garden of her own, but she was a gardener who could not recognize the plants in her own yard. Obsessive by nature or by necessity, she would compensate by doing all she could to keep that garden fresh in her memory.

"Instead of counting sheep at bedtime," Janice told me, "I would walk, or rather 'talk,' through my garden in my mind." She would repeatedly visit each plant, say its name, list its features, then move on to the next one. Her nightly ritual might sound something like this: "Back by the fence there is a clematis that blooms in big purple flowers, and in front of that there is a bunch of white coneflowers." She could never call up an image of her garden or even individual plants, so she relied on associative language to help her remember them.

For a long time, volunteering with the National Ballet of Canada was an all-consuming passion for Janice. She immersed herself in the history of the company and knew every dancer—from the prima ballerina to the latest intern in the corps de ballet. When *Swan Lake* opened, she went to all twenty-three performances, sitting in the front row. Yet she could not tell the dancers apart as she viewed them on stage or when she mingled with them later. Eventually she devised compensations. One was to memorize the program so she would know who was playing what character on any given night, and the other was to study their ways of moving until she spotted idiosyncrasies—this one's long limbs or that one's stiffness in the hips that led to a certain way of moving.

In her book *Visual Agnosia: Disorders of Object Recognition and What They Tell Us about Normal Vision,* Martha Farah notes that one set of cortical areas perceives static visual cues and another motion-based cues, so we can separate form from motion when using cues to identify people. This is how Janice used characteristic movements to distinguish one dancer from another.

Janice once took a course in guided meditation. The students were asked to imagine themselves in a forest clearing and to follow a path to a garden gate before encountering someone on a bench—a feat her fellow students accomplished easily. Janice was acerbic and funny as she described what happened, or didn't happen, to her: "Everybody was saying 'Oh my God, wasn't that amazing? I met my mother, who's been lost for years . . . I met my spirit guide.' This and that and the other. And I hadn't met a damn person or creature or anything. I never got as far as the garden gate. So I never went back to that group."

Janice has found that people have difficulty believing her when she says that all she sees when she closes her eyes and tries to conjure up an image is total blackness.

Research, especially Stephen Kosslyn's work at Harvard, has demonstrated considerable overlap between the neural networks used for perceiving visual stimuli such as objects and faces and those used when forming a mental image of objects and faces. This would explain Janice's difficulty with forming mental images when meditating. The same cortical areas resulting in her impairment in perception and recognition are those necessary for generating the mental images required in her meditation exercise.

Janice has thought deeply about what it is to have neurological weaknesses, and she has made peace with herself. But there is no denying the hurt and the heartache. "I feel really good about myself and my life, and how I've compensated for any deficits that I have," she told me. "I've made terrible mistakes in every area of my life, as we all have. I don't think I did any worse with what I was given than what most do with what they're given. I feel happy to be who I am. And I forgive myself for hurting people's feelings when I don't recognize them. But I still find it difficult."

What is going on, or, more correctly, what is *not* going on in the brain of someone with an object recognition weakness? Oliver Sacks offered a possible answer to that question in another *New Yorker* piece, this one entitled "A Man of Letters" published in June 2010. In that essay, Sacks suggests the following: "It would be uneconomical to suppose that there are individual representations, or engrams, for each of the billions of objects around us. The power of combination must be called on; one needs a finite set or vocabulary of shapes that can be combined in an infinite number of ways."

This "vocabulary of shapes" may be limited in someone with object or face recognition problems. An impaired ability to initially perceive specific features of objects or faces perhaps leads to a reduced or imprecise stored image bank of such features (whether an exact match or just the defining elements such as shape, contour, or surfaces). This would mean that each time that person encounters a particular object (say, a jar of mustard) or a face (his mother's), there is an impoverished frame of reference to aid in efficient recognition. As a result, the individual both recognizes less and remembers less.

To understand this neurological deficit, think of a familiar animal, a cat. Now close your eyes and call up an image of a cat. Can you see the cat in detail—ears, tail, markings, body—or is the image fragmented, with only individual details appearing? Is it clear or is it fuzzy? Those with exceptional functioning in this area see the cat in detail, as if looking at a photograph. Those with very poor functioning see a blank screen with no image.

Perception and recognition of objects or faces and symbols are parallel processes: the right hemisphere's occipital-temporal network governs objects and faces, and the same network in the left hemisphere governs symbols (letters, words). People who have a problem in this left hemisphere network struggle with reading or spelling (which I discuss in Chapter 13).

Interestingly, it is thought that these networks in both the right and left hemispheres were originally involved in object and face recognition. When humans developed written language and began to read, the plastic brain had to adapt, recruiting areas that had evolved for purposes other than reading. The left hemisphere's occipital-temporal region was recruited to handle the symbols of written language, while the right-hemisphere area remained specialized for objects and faces.

Since 1980, I have called the function of the cognitive area in the right hemisphere *object recognition*, and at that time it was thought that faces were just another form of object. Even though we now know that one network in the right hemisphere enables us to recognize objects, and another partially overlapping network governs face recognition, for continuity's sake I have chosen to retain the name *object recognition*. Research is under way to establish each cortical area's function and to determine precisely when they are active during the object or face perception process, which will help determine how these networks actually work.

Navigating our world relies in part on our ability to recognize objects. High spatial ability orients us in space and allows us to visualize our path and construct maps in our head; object recognition allows us to fix in our minds particular landmarks and use them as guides. Someone with good object recognition will always be able to retrace his or her steps using these landmarks as signposts.

An individual with a problem in this area of the brain takes longer to visually recognize and locate objects. Those with a severe weakness in this area are almost object-blind. Such a person cannot sort large spoons from small spoons but sees only a collection of spoons. One young man we treated had such low object recognition functioning when he came to us that he couldn't distinguish between the cupboard door and the front door, and he often walked into the former when he meant to use the latter. When he started at the school, he was unable to go to his shelf and retrieve the books he needed for the next class because none of the normal visual aids (the shape, color, and size of a book) could help him. And because he couldn't recognize his personal belongings, even finding his shelf was impossible.

Object recognition is an essential strength for artists, detectives, archaeologists, and pathologists. If you're strong in this area, you're adept at putting jigsaw puzzles together, you could recognize as a child any make and model of car even at a glance, and you're the member of the family who can always turn up what someone else has lost. You excel at "Where's Waldo?" and "I Spy" games.

But if you have an object recognition deficit, you attempt to enter someone else's car in the parking lot or pick up someone else's coat at a party, and when you put something down in a room, even though it's there in plain sight, you can't find it. I sometimes call the object recognition weakness "the refrigerator dysfunction." The classic example is the teenage boy (men seem to have this problem more than women) with the fridge open and even though he's staring right at the jar, he shouts out, "Mom, where's the mustard?"

For pathologists, object recognition is a matter of life and death. A young doctor came to us after an incident, which he describes:

My new position as a pathology resident was unexpectedly difficult. The job required surgical skill, rapid visual perception, and quick judgments. One day, I was determining whether surgeons had completely removed a breast tumor. In order to mark the anatomic orientation of the tumor resection, I used different colored inks on the borders of the specimen. This delicate work took great effort and all my concentration.

I was proud of my careful work and I understood that the patient's life might depend on me doing my job right. When the supervising physician stopped by to check my work, I briefly and accurately summarized the patient's clinical history. She seemed pleased until she looked at the specimen.

"Don't you see that you've mislabeled the anterior margin as superior?" she said. "The anterior margin is folded over the superior margin!" Instantly, I saw what she meant. There was a fold of skin as big as a silver dollar hanging over the superior edge of the resection, which seemed to materialize out of nowhere. In half an hour of carefully examining the tissue, I had not seen this, yet after it was pointed out I could see it quite clearly. I wondered what was wrong with my vision.

After a long diagnostic process, I found that I had a nonverbal learning disability with weakness in visual perception.

He temporarily withdrew from his residency program and enrolled in the Arrowsmith Program before taking up his medical training once again.

The object recognition exercise, like all the others in the Arrowsmith armamentarium, has a history. In 1980, a twenty-one-year-old art student came to us looking for help. She had all the capacities necessary to become a successful illustrator save one: object recognition.

In terms of overall intelligence, she ranked in the 99th percentile, but on her ability to remember visual details of objects, she was in the 25th percentile. This meant that it took her longer to complete drawings because she had to pay careful attention to ensure she didn't miss any details—a button on a coat, say, or whiskers on a cat.

This young woman had an uncommon sense of color and excellent hand-eye coordination, but she was always losing things—placing brushes on a table, for instance, and then not being able to find them. If she put her gray tweezers down among pens and markers of the same color, she would be a long time trying to find them again.

In order to stimulate the weak parts of her brain, I developed an exercise in which she would have to memorize a particular image and then pick it out from a display of very similar images. Fine discriminations among a wide range of visual features are required to carry out the task. The varied stimulation in the exercise targets the different functional aspects of the cortical areas involved in the object and facial recognition networks. As people do the exercise, their ability to recognize and remember both objects and faces improves.

The art student improved, and she no longer left out details in illustrations and had no difficulty finding objects. Several years after she had completed the program, I saw an article about an award she had won for her work. She is now a successful artist and illustrator.

Annette Goodman is a cherished colleague and uniquely positioned to talk about neurological deficits: she knows the territory intimately from both theoretical and practical standpoints.

This is Annette's life as viewed through the window of the object recognition deficit.

She likes a bare and spartan house so everything is in plain sight. In every closet, items are lined up like little soldiers so she can memorize where every object is. Annette keeps family photos in an accordion folder labeled by year to help her identify who is who—and heaven forbid if someone messes with the order.

At a movie, she is constantly asking the friend next to her, "Is that the good guy or the bad guy?" Friends can't fathom that three-quarters of the way into a movie she still can't discriminate between the main characters if they are both the same gender.

When Annette runs three-week training courses, it takes her a full week to remember the face that goes with each name. To retain recognition throughout the course, she silently goes up and down the rows while repeating the names to herself throughout the day. A month later, the faces and names are all gone.

Unable to see the visual details of artwork or sculptures, she politely declines when her friends invite her to go on a museum trip. For years, her husband could not convince her to go on a vacation. Why go to Paris and Rome when the images will only disappear? Her memory-for-information deficit compounds the problem: she does not remember any of the information she learns on museum or walking tours.

Annette has three children, and she had no answer to that oft asked question after every birth: Who does your baby look like? She finds grocery shopping—indeed all shopping—stressful. Without visual images in her head (a favorite can of pickles or box of cereal), she can't do the quick matching that most people find second nature. As a result, she orders her groceries and has them delivered.

Annette went shopping for birthday wrapping paper. She found one with triangles sprinkled throughout and showed it to her daughter. The little girl pointed at the triangles and said, "*Ima* [Ma in Hebrew], those are Christmas trees!"

Annette will meet a friend not seen for a while, and the friend might say, "Don't I look great? I lost ten pounds." Annette has to apologize for not having noticed, and she worries about appearing insensitive or self-absorbed. There is no before image in her mind's eye to compare with the after.

Annette knows Janice Mawhinney, and recently they decided to do the cognitive exercises together to overcome their object recognition deficits. Janice is making good progress—evident in a note she sent me in December 2010:

While I am not seeing much, if anything, while I do my exercises, I did have a surprising thing happen this morning. I had gone to the Whole Life Expo on the weekend and had a wonderful reflexology treatment from a very kind woman who advised me on a special cream, some foot exercises, and a book that could help with my ankle and knee pain. The cream and exercises did help some, and I have just received the book I ordered on her advice. I wrote her association to thank her; they passed my e-mail along to her and I heard directly from her by e-mail this morning. I then looked up her name on Facebook. It is a very common name, and there are pages and pages of women who have that name. However, I instantly recognized her photo when I saw it.

How could I possibly recognize her photo after such a brief time with her in a crowd last weekend, when I repeatedly walked past people in the *Star* hallways after working with them for many years, not recognizing whether they were anyone I knew? I can only think that even though I do not appear to be making progress during my exercises, something is shifting somehow in my brain in a very positive way.

Annette also noticed significant change over the course of several months. Three weeks after beginning, she noticed the first sign: the image of an artichoke popped into her head. She could see the luscious vegetable from above with all its petals, and she could imagine herself peeling off the petals one at a time, dipping them in mustard sauce, and eating them. Janice suggested that the artichoke be her good luck charm: the first image ever to appear in Annette's head.

The changes kept coming. For the first time in Annette's life, visual images of objects were making a memorable impression. An illustration of a gem on a birthday card. Yellow and green spools of cable wire hanging from a pole on her street. Sharp details of clothing worn by a speaker at a conference. Furthermore, Annette was able to recognize what the abstract art her daughter had painted was meant to represent.

As for faces, she started recognizing acquaintances even after new haircuts or glasses. She surprised herself: she met a woman who looked, she thought, like a neighbor, and it turned out that the two women in question are sisters. Annette is now confident that she will recognize the facial features of her grandchildren when they are newborns.

Annette is now shopping for a wedding gown for her daughter and can go to multiple stores, watch her daughter try on countless gowns, and later discuss the pros and cons of each gown. "I can be a full participant in this very special shopping experience. It is very sweet," Annette says. "I feel like a mother."

Two unexpected bonuses came out of all this work. One is that in addition to being able to call up images (dishes ordered in restaurants, a scene from a recent social gathering), Annette is now seeing images from the more distant past—images that must have somehow encoded themselves prior to her starting the exercises. She recently called up images of colleagues she had worked with briefly several years ago, and she also remembered a childhood blanket she loved—thin red wool with a white pinstripe. She could even see its texture. And second, Annette is having an easier time remembering names, even with her memory-for-information deficit so far unaddressed (but not for long). Given Annette's previous combination of deficits, she could not call up both faces and names without a struggle. Now her strength in the object recognition area is supporting the memory function for names.

Ten months after starting the program, Annette decided to retest on the Woodcock-Johnson Picture Recognition test: she had improved from the 6th percentile to the 88th. And while the numbers are impressive, what is even more exciting to Annette is that when thoughts pop into her brain, pictures now accompany those thoughts.

CHAPTER THIRTEEN

A CLOSED BOOK

Once you learn to read, you will be forever free.

—FREDERICK DOUGLASS *(1818–1895)*

Jeremy Johnson grew up in the Haliburton Highlands, northwest of Toronto. The son of two teachers, he would tell them when he was very little that his inability to remember words and letters was not a problem because "my wife will remember." But as he got older, he began to worry that his learning disability would get in the way of meeting girls (never mind marriage), and his cavalier attitude soon turned into anguish.

"I could not read at all," Jeremy recalls. "I could not put ideas on paper. I could not organize my thoughts. I was pretty much a train wreck."

Jeremy would copy letters as the teacher wrote on the blackboard, but he could never keep pace, and he wrote each letter of the alphabet without identifying what the collected letters meant as words. He would still be writing down from the first blackboard and the teacher was on to the third; when the teacher began to erase that first one, he'd panic. A writing assignment that took others in the class half an hour took him eight times as long.

"The scariest thing for a learning-disabled kid," Jeremy told me, "is a substitute teacher. Sometimes they ask students to take turns reading aloud from a book. On a break, I'd tell the teacher I could not read. Or I'd go to the bathroom. I'd make an excuse."

In a restaurant, he could not read menus. He'd order what the others were having or, in a fast food restaurant with pictures above the cashier, he'd look up and order "Combo #3" (Jeremy had heard his friends say the word and he could recognize individual numbers). Which restroom should he use? Even as a teenager, he would have to ask his mother when there were no images, only words, on the restroom door.

When a written joke was passed around a classroom, Jeremy would time the response of the others so he could gauge how long to "read" the joke before laughing. In a movie, he couldn't read subtitles. Driving a car, he couldn't read street signs.

Written language evolved as a way to represent spoken language, using symbols to represent sounds. The written representation is arbitrary in the sense that it was agreed on as a set of rules by the speakers of that language. Some written languages, such as Italian, come very close to having a dedicated symbol or sequence of symbols for each phoneme (or distinctive speech sound). These are known as phonetic languages.

Italian has thirty-three letter combinations for its twenty-five phonemes. English has 1,120 different ways of spelling its more than forty phonemes, making it one of the least phonetic written languages. Think of the *f* sound in *tough* or *cuff* or the *au* sound in *bough* or *bow*. Italian is a much easier language to learn to read than English; in Italy, the reported rate of dyslexia is half that of the United States.

What exactly is dyslexia? Sally E. Shaywitz and Bennett A. Shaywitz, codirectors of the Yale Center for Dyslexia and Creativity, state that "developmental dyslexia is characterized by an unexpected difficulty in reading in children and adults who otherwise possess the intelligence and motivation considered necessary for accurate and fluent reading." However, no two children with a reading disorder present exactly the same in the classroom. A number of cognitive areas work together to carry out the task, first of learning to read and, later, of reading fluently. An impairment in any one of these areas will interfere with the reading process in a way that is particular to the function of that area. Functional magnetic resonance imaging (fMRI) technology can now pinpoint many of the areas of the brain activated by the process of reading.

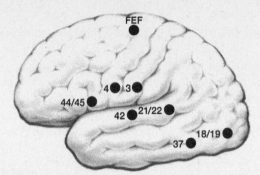

Figure 4. The Brain and Reading.
(Adapted from Lassen et al., 1978; Fiez and Petersen, 1998; Dehaene, 2009.)

Brodmann Area	Name of Area	Lobe	Role in Reading
44/45	Broca's Area	Frontal	Speech production, phoneme articulation, phonological processing in reading necessary for sounding out words
6	Frontal Eye Field (FEF)	Frontal	Involved in eye movements when reading
4	Primary Motor Area	Frontal	Controls movement of the mouth to say the words when reading out loud
3	Somatosensory Area	Parietal	Provides sensory feedback to mouth movements for clear articulation when reading out loud
42	Primary Auditory Cortex	Temporal	Hearing of sounds—pitch, frequency, volume
22/21	Superior and Middle Temporal Regions	Temporal	Discriminates speech sounds from other sounds, perceives and identifies phonemes (sound units of language) necessary to understand language, beginning of phonological processing, early part of process to attach sounds to letters, lexical memory for words
18/19	Secondary and Visual Associative Areas	Occipital	Early stage of processing visual form of letters
37	Occipito-temporal Region, Fusiform Gyrus	Temporal	Visual word form area/brain's letterbox/symbol recognition—visually identifies letters and words

In this chapter we will be focusing on three cognitive areas involved in learning to read English: the first is the symbol recognition area, required to learn the visual representation of each letter in the alphabet and later to learn to recognize words—in other words, to build a sight word vocabulary. Broca's area in the brain (described more fully later in the chapter) is a component in phonological processing. This part of the brain allows the individual to learn sound-symbol correspondence (converting letters into sounds) and then to sound out new words and learn their pronunciation. Third, and equally important, the motor symbol sequencing area allows one to track words smoothly across the page when reading. A deficit in any one of these three areas will interfere with a student's ability to read English; a student with a significant deficit in all three areas will not be able to learn to read beyond an elementary school level.

A fourth area of the brain, the superior temporal area in the left hemisphere (covered in more detail in Chapter 21), requires brief mention. One function of this area is to discriminate speech sounds from other sounds and to identify phonemes. In reading, this area of the brain plays a role when the child begins to match the sound (phoneme) to the written letter (grapheme). Sound-symbol correspondence, as it's called, is an early stage in phonological processing. Originally it was thought that Broca's area carried out a significant part of this function. Ongoing research is investigating the precise role of these two areas in the phonological aspect of reading. What is clear is that both areas make important contributions to translating the printed word into speech sounds.

The exercise I developed to address the functions associated with Broca's area requires that the student discriminate between phonemes. This is what we see as an outcome of working on the Broca's exercise: improved auditory speech discrimination, an ability to learn sound-symbol correspondence, improvements in speech, and improvements in reading. In this chapter, I have included sound-symbol correspondence as a feature of the functioning of Broca's area because it does play a role. Implicit is the understanding that the superior temporal area also plays a role in this process.

In his book *Reading in the Brain*, French neuroscientist Stanislas Dehaene describes a process I call *symbol recognition*: how the human eye first registers a number, letter, or word on a page and, in less than one-fifth of a second, transmits that information to the occipital-temporal area in the left hemisphere of the brain ("the brain's letterbox"). This area, also referred to as the *visual word form area*, recognizes and remembers the visual pattern of symbols that make up words. This area then passes the information to other areas of the brain required for reading.

Years ago, a problem in this cognitive area was referred to as "word blindness"—the person looks at a word but cannot recognize the letters or retain the visual image of words. To someone with word blindness, print on a page looks like meaningless squiggles.

A symbol recognition weakness impairs the ability to read. But if you have a strength in this area of the brain, you can do all or some of the following. You have a photographic memory for text and can easily do word searches. You can read text when it's upside down because you can recognize words regardless of orientation. You can read messy handwriting or scrambled words (because your ability to hold the image of words allows for matching—even when individual letters are poorly formed or in the wrong order).

In Chapter 12, on object recognition, you called up the image of a cat. Now think of the word *cat*. Close your eyes, and visualize a blackboard. Can you call up an image of the word? If this is easy, think of a multisyllabic word and repeat the process. Can you see the image? People with this deficit may see no image, a fuzzy and incomplete image, or an image that quickly disintegrates. With good functioning here, you can see and retain a clear image.

In 1861, Pierre Paul Broca, a French anatomist, observed that the ability to speak was lost following damage to a certain spot in the left inferior frontal cortex of the brain, and this region was eventually named after him. Subsequent research has reinforced the language-central role of Broca's area, but there is debate about its various functions. Researchers now suggest that since it appears to be composed of numerous functional subunits, Broca's area should be referred to more accurately as Broca's complex. I am focusing here on the role of Broca's area of the brain in the phonological aspect of the reading process.

An English speaker with a severe deficit in Broca's area approaches the task of pronunciation when reading with the same befuddlement that strikes most of us when we look at Welsh words. Imagine having to read this Welsh word: *Llanfairpwllgwyngyllgogerychwyrndrobwllllantysiliogogogoch*. It means "St. Mary's church in the hollow of the white hazel near to the rapid whirlpool and the church of St. Tysilio of the red cave." English readers would find these letter combinations unusual and the word impossible to pronounce without knowing the rules in Welsh for sounding out letters and letter combinations. We stare at the word feeling helpless, not knowing where to start. And even if we make a stab at it, we have no idea if we have pronounced the word correctly. This is how those with Broca's deficit feel when they are presented with a word never encountered before. They have trouble learning the rules for pronunciation and so cannot call on those rules to break down the word into its component sound units, sound these out, and then blend them into a word.

A student with a deficit in Broca's area and a strength in both the symbol recognition and motor symbol sequencing cognitive areas will eventually learn to read English as a host of sight words and will experience difficulty only when trying to learn the pronunciation of new words. These individuals frequently learn to read well, albeit a bit later than their peers (as a result, it is often thought that they have "matured"). Since the underlying cognitive problem has not been addressed, these students will continue to struggle with the pronunciation of new and complex words. Their oral or spoken vocabulary may be smaller than their silent-reading vocabulary, and their speech expression may be affected.

For most of us, reading is like breathing. We do both without thinking (or so it seems), and both activities are life sustaining. Anyone who loves books cannot imagine a life devoid of reading. The great pity is that when reading does not come easily to a child, that state of affairs will not likely change without cognitive intervention. Two American authorities on childhood development, G. Reid Lyon and Louisa C. Moats, did a study in 1997 demonstrating that 74 percent of children who are poor readers in the third grade remain poor readers in the ninth grade.

Jeremy heard about Arrowsmith in tenth grade. He was then operating a little landscaping business, tending neighbors' lawns and gardens, and one of his customers told him about our program in Toronto.

"Grade 10 was the breaking point," Jeremy recalls. "School was more intense now, with more reading and writing. My mother had no time for her own teaching. She was taking so much time with me that I worried she'd lose her job."

Two weeks before starting at Arrowsmith, Jeremy was tested. One of the staff, Andrea Peirson, will never forget it. "I was almost crying," she says, "seeing the discrepancy between his ability to read and write, which was practically nonexistent, and his visual memory for objects, which was off the chart." Jeremy, with his highly developed right hemisphere, thought in pictures. An area in the left hemisphere of his brain, however, was not responding, and his visual memory for words was severely limited.

In fact, Jeremy had significant deficits in all three areas related to the mechanical aspect of reading. He had a symbol recognition deficit, which made learning sight words close to impossible. He had a Broca's problem, so the ability to link sounds and symbols was impaired and, with it, the ability to use phonics to sound out words. And his motor symbol sequencing deficit had an impact on his ability to smoothly track letters, words, and punctuation while reading. No wonder reading eluded him: the various parts of his brain that enable these skills were all underperforming.

The Arrowsmith exercise designed to address the symbol recognition deficit involves the study of languages unfamiliar to the student, such as Arabic or Urdu. The aim is not to speak, read, or write the language but to become proficient at recognizing the shapes of the letters and symbols. Jeremy would spend time every day locking onto one particular image and then would try to hold it in his mind's eye, as though it were an image developing on a photographic plate. Then onto the next one, maybe a trickier one, to hold and develop. The name of the game is a stronger visual memory.

Fourteen months into the program, when he was seventeen years old, Jeremy saw a major change. His brain was performing significantly better in the key areas related to reading. As a result of working on the symbol recognition exercise, Jeremy could now retain the visual images of words in memory. He could quickly memorize the look of a word and later recognize it when presented with that word in another context, so he was now able to build a sight word vocabulary that was age appropriate. And because he had worked on the cognitive exercises for his Broca's and motor symbol sequencing problems, there were other improvements. Jeremy's brain could now remember the sounds of the letters and phoneme units that make up words, meaning that he could sound out unfamiliar words as an additional aid to build his reading vocabulary. Finally, Jeremy's brain was able to learn the motor plan required to track symbols in a sequence, thus enabling him to read quickly and efficiently without having to use compensatory strategies such as a ruler or finger on a line of text.

At home that Christmas, he was able to read aloud the nativity story from the Bible; his mother cried as she listened. Not long afterward, he read his first novel: *To Kill a Mockingbird*. It took him several hours to read a few pages, but by the end of the book he was cruising along at eight pages an hour—for him, the speed of light. "It was amazing," Jeremy remembers. "Watching myself increase in speed and understanding. It was so exciting. Clearly the exercises were working to allow me to read."

Today Jeremy lives in Calgary with his wife, Lois, and is a mechanical engineer working as a maintenance analyst for pipelines used in the oil and gas industry. Jeremy always had a strength in spatial, mechanical, and abstract reasoning, and after addressing his reading difficulties, he was able to pursue a career that put his above-average right hemisphere strengths to work.

The Arrowsmith School in Peterborough is located on the northern outskirts of my old hometown. Upstairs in the boardroom, we meet with Jill Marcinkowski, the assistant director, and her oldest student, Marcel Peeters.

"I'm forty," Marcel says. "I should be looking for my midlife crisis." Instead, Marcel is working on cognitive programs to address his severe difficulties with reading, spelling, and writing. It is no surprise that Marcel's spelling is poor: learning to spell and learning to read place similar demands on the brain.

Marcel cannot spell his own middle name and uses his brother's mailing address (the two live close by) because Marcel cannot spell his own address. For a time he worked for his brother, but that meant looking up customers' names in the telephone book, which took him an interminably long time, so Marcel would dial the operator, incurring charges.

Marcel took six years to get a license to sell mutual funds and insurance. Someone so unencumbered would have obtained that certificate in less than a year.

During his brief career as a financial planner, Marcel wanted to send all his clients a Christmas card. He came up with, "Wishing you health and happiness in the New Year. Sincerely, Marcel." And as he always did with anything he intended to mail, he typed the salutation on the computer, hit SpellCheck, got the verification he needed, and then handwrote a great many cards.

"Well, somehow," he said, "I spelled *cynically* instead of *sincerely* and the computer said it was spelled right. So I sent out all these Christmas

Lyova Zazetsky and Aleksandr Romanovich Luria.
Reprinted with permission from www.descopera.org

1951: Young family home, 301 Bogert Avenue, Toronto.

Winter 1952: Louie May Arrowsmith Young
with granddaughter Barbara, Toronto.

October 1952: Barbara in backyard, Toronto.

Summer 1954: Donald and Barbara, front porch, Toronto.

July 1954: Greg, Alex, and Barbara, front porch, Toronto.

Summer 1955: Barbara and Donald with father, John, Lake Simcoe.

Christmas 1955: Barbara, Greg, Alex and Donald, Toronto.

Spring 1956: Donald and Barbara with grandmother
Louie May Arrowsmith Young, Toronto.

Summer 1956: Building Muskoka cottage—Alex, Barbara, and Mary Young.

November 1956: Donald, Alex, Mary, William (*in mother's arms*), Barbara, and Greg, Toronto.

Summer 1957: Donald and Barbara, Young's Point, Ontario.

1960: Barbara, Peterborough, Ontario.

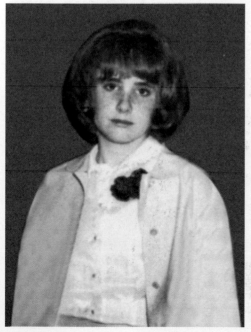

June 1965: Grade 8 graduation,
Queen Mary School, Peterborough, Ontario.

Fall 1965: Barbara with guinea pig and Mr. Cat/Star.

Fall 1966: Grade 10.

1969: Barbara with her brothers Greg, Alex, and William (*in striped shirt in front*), Peterborough, Ontario.

Fall 1969: Sewing.

Duke of Edinburgh's Award certificate.

October 1969: Barbara receiving Gold Duke of Edinburgh's Award from Prince Philip, Peterborough, Ontarlo.

1972: Barbara with Mr. Cat/Star.

1974: Graduation, University of Guelph.

Wedding day, August 30, 1980: Barbara with her mother and father,
Mary and John Young, Peterborough, Ontario.

Wedding day, August 30, 1980: Alex, Donald, John, Mary Young;
Joshua Cohen; Barbara, Greg, Anne, and William Young.

June 1995: Arrowsmith School graduation, Toronto.

2000: Barbara in her office, Arrowsmith School, Toronto.

2000: Arrowsmith School, Toronto.

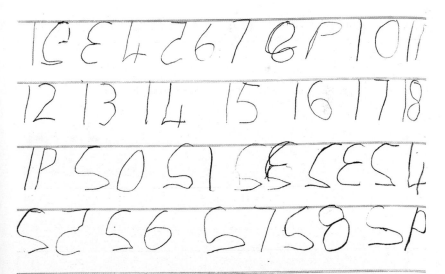

Grade 1: Number sequence 1 to 29 with multiple number reversals.

Grade 1: Counting by tens showing number reversals.

Grade 1: Counting by tens, 39 to 100, showing number reversals.

Grade 5: Two-digit-column addition questions attempted twice and answers incorrect both times.

Grade 5: Three-digit-column addition questions attempted twice. Four subtraction questions attempted twice with only one correct answer. Teacher comment: "Time wasted."

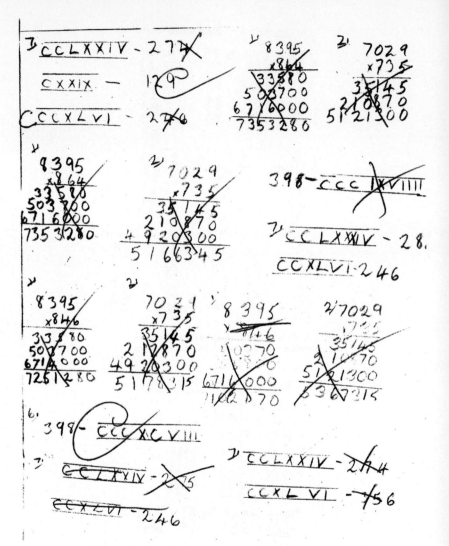

Grade 5: Roman numeral questions—six questions, multiple attempts, two correct.
Two multiplication questions attempted three times, none correct.

cards that said, 'Wishing you health and happiness in the New Year. Cynically, Marcel.'" That and another faux pas spelled the end of his career as a financial planner.

In a doctor's office, he'd be asked to fill out a form—and couldn't. "Get a computer you can talk to," he was advised. In other words, use compensatory devices. "But it's just not realistic," Marcel says. "When you're at a doctor's office, what are you going to do—pop open your laptop?

"It's been a real struggle," he concedes. "I was doing financial planning but not making any money at it. At night I worked in a group home, and on weekends I sold vegetables at a market. So I was working over a hundred hours a week."

One bit of good news is that a year of cognitive exercises has allowed him to jump three grade levels in his reading skills—from seventh to tenth grade. When I saw him, he had read three books in the past ten days, including *Freakonomics*.

Meanwhile, Marcel has started to write his own book. Jill Marcinkowski helped him come up with a title: *Cynically Yours*.

People with learning disabilities are often told, even reassured, that adaptive technologies will allow them to get around the problem of reading and writing. And while some of this compensatory software is marvelous, it's not foolproof. Just ask the learning-disabled John Jackson, who knows the territory.

A twenty-five-year-old of Celtic lineage, he was raised in a small town north of Toronto. When I talked to him in October 2010, he had just started a one-year full-time program at the Toronto Arrowsmith School—after completing three years of university courses. John had graduated from high school and made the honor roll three years in a row. That brief sketch would suggest a rosy picture, but a deeper look reveals another story—and the many flaws and limitations of compensatory technology.

At university, John took less than a full course load each year while latching on to multiple compensatory technologies, including one that scanned textbooks and then read them out loud, voice recognition software to get around writing, and a digital recorder for note taking. Each technology had an unforeseen flaw. The first device deployed a computer-generated voice, which drove John to distraction. The voice recognition software went off the rails when it heard words such as "The Aeneid" or "Fenian Raids." And the recorder worked fine until a professor who had lived under a dictatorship in his homeland and was fearful of being taped forbade the device in class.

John was, as he puts it, "basically born dead. I was blue, and they had to revive me. I had my umbilical cord wrapped around my neck blocking oxygen to the brain." Medical authorities later diagnosed John as having mild cerebral palsy, which they believe led to his learning disabilities. Once that assessment was finally made when he was fourteen, John was given the sort of help that learning-disabled students typically get: extra time on tests, a scribe, a digital recorder, and devices that convert text to speech. But even with all this human and technical help, John still had to rely heavily on his parents to proofread all his essays.

"I have my parents proofread my e-mail and Facebook messages before I send them since I don't trust my proofreading skills," John says. It means he has no privacy, but he sees no alternative. "In my first year in university, I was seeing a girl I really liked; we had mainly been communicating by telephone and sometimes by Facebook. When a close friend of hers died, I sent her a message on Facebook just asking how she was doing. I got an angry message back. I had not realized that in the last few e-mails, I had spelled her name wrong. I had sent her messages using voice recognition software. What I had been saying into the microphone was not what the program was interpreting. I had never felt so ashamed in my life. It took weeks before she spoke to me." They are friends now, but John's hopes for something beyond friendship have been dashed.

We had asked John before coming to Arrowsmith to write some reflections on his life and school experience to that point. Clearly, it's been a roller coaster, for this is what he wrote:

> My academic journey has been a rough one. In elementary school, I was held back in grade 1, which led to me being teased and bullied by other children. I was encouraged to use a computer for school work, which made very little difference because SpellCheck could not find the words I was looking for. In grade 7, I was assigned an educational assistant (EA) for a few school days a week. My marks improved so dramatically that the school felt I no longer needed an EA for grade 8.

John graduated from high school with very good grades, but he is at pains to stress that without his parents transcribing all his work and helping with two hours of homework every night, he would never have graduated. He thinks the diploma should have an asterisk beside it because his reading, writing, spelling and grammar are all highly suspect. He says:

> Learning problems have been a source of much of my frustrations all my life. I become easily stressed out over tasks that the average person can do. For example, writing a simple phone message down is a challenge unless I ask the caller to spell his or her name slowly. Those moments often leave me feeling depressed about myself and my outlook on life. I often think that no woman would want me, let alone marry me or raise a family with me.
>
> There is a stigma if you can't write clearly or spell correctly. People assume that you're not smart or socially interesting. Because of my learning problems, I've always felt socially awkward. Especially in social settings that may involve writing. I am always afraid of a party game like Pictionary, Charades, or any game where you have to write the answers.

John Jackson will not feel worthy of his high school diploma and university degree until he can do what his peers can do unassisted: spell easily, read and write well, and pen a note that will not raise an eyebrow or cause its author acute embarrassment.

Given the function of Broca's area of the brain, a student with this deficit will struggle to learn the spoken aspect of a foreign language. Foreign words are just sounds initially, with no meaning attached to them; they need to be discriminated and then remembered and said as phonemic units. As noted earlier, cognitive areas in the superior temporal lobe (necessary for phoneme recognition) contribute to this process along with Broca's area. With repeated exposure and immersion, most students learn the spoken aspect of their native tongue, regardless of the language or any learning disabilities. Learning to speak a second language, though, will be difficult with a Broca's deficit.

Many students in the Arrowsmith Program are enrolled in Jewish day schools and required to learn Hebrew, which is more phonetically complex than English. Students with a Broca's deficit therefore experience difficulty not only with the spoken aspect of Hebrew but with the mechanics of reading it as well.

A little knowledge of how Hebrew works will illustrate the depth of the challenge. The majority of written words in Hebrew are based on three consonant root words. Vowels rarely appear in print to aid in pronunciation. For each printed word, there are multiple possible pronunciations, depending on which word is intended in a particular context. Consider for example the following Hebrew words (all transliterated into the Roman alphabet). The printed word *KTV* (to do with writing) may be read as *KoTeV* (writing or author), *KaTTav* (reporter), *KaTuV* (written), *K'TuVim* (scriptures), *KaTaVti* (I wrote), *K'ToVet* (inscription or address), and so on. All of this places a greater load on Broca's area because readers are required to quickly generate many possible pronunciations.

For Michoel Fixler, learning Hebrew was a huge struggle. He had a strength in the symbol recognition cognitive area, but a Broca's deficit meant he could read English in third grade but not Hebrew. Because the underlying cause of a reading difficulty varies from one student to another and because different cognitive areas become important for reading in different languages, a student may be considered dyslexic in one language but not another.

"Don't make me read," Michoel would beg the rabbi hired to tutor him in Hebrew. Teachers tried to assure his mother, Reva, that with time and practice, he would learn the language.

"I had Michoel pulled out of class for years," she recalls. "He had a special reading teacher, and the rabbi read with him every day, and they kept saying, 'Practice, practice, practice,' and we did practice. I terrorized the child. It was bad; it was really, really bad. Michoel was being tortured; I was being tortured. We were fighting like crazy." And after all that, Michoel's grades remained in the 50s. It was impossible for him to remember the 250 root words he might be tested on during any given exam.

That was six years ago. In the wake of several years in an Arrowsmith classroom, here are Michoel's eleventh-grade marks in language subjects: in Spanish, 90 percent, and in his three Hebrew subjects, between 85 and 95 percent.

The Arrowsmith exercise to stimulate the Broca's area of his brain mirrors the cognitive demands made by reading Hebrew: Michoel was asked to listen to sounds—the sounds that make up words. His task was to hold onto those sounds, which began simply and grew more complex. Michoel would be challenged to repeat the sounds, then play with them, with the emphasis here and now there. This rapid switching placed a load on the Broca's area of his brain, resulting in a greater facility in learning languages.

A significant weakness in motor symbol sequencing makes reading a struggle—not because the student can't learn sight words but because he or she will misread sight words (even ones they know) due to an eye tracking problem. With this deficit, the eyes do not track smoothly across the page and, as a result, one could read *horse* for *house* or *cat* for *car*. A student with this deficit will labor well beyond the initial learning process since smooth eye tracking is always needed to read.

Reading English requires a significant amount of eye tracking, and reading Hebrew places an even greater load on the motor symbol sequencing area. In Hebrew, the reader needs to track both the letters and the small dots underneath the letters, which represent vowel sounds and are present when students start learning to read Hebrew. Poor tracking makes reading English difficult; it makes reading Hebrew a herculean task.

Of all the learning disabilities addressed by Arrowsmith, difficulties with the mechanical aspect of reading take the longest to deal with. Reading takes more time than other academic subjects, as educators say, to "bring to grade level." The problem is twofold. First, it may take several years to address all three areas related to reading, so students can have reading deficits even as they move through the program (though these are progressively less pronounced over time). And, second, these students have been unable to benefit from reading instruction and exposure to print, so there is much they need to learn. In addition, students with difficulty reading may develop a deep-seated aversion to it. After they've finished their course at Arrowsmith, they may require an additional year to overcome that antipathy and start to read for pleasure. In order to close this gap sooner, Arrowsmith staff often work with resource support staff, with the latter teaching content and reading and the former addressing the cognitive problems.

Avital Goodman, Annette Goodman's daughter, needed three years to complete the three exercises related to reading, and only at the end of her third year at Arrowsmith was she reading at grade level. At the end of second grade, Avital was rated in the 4th percentile in reading on the Stanford Achievement Test, and at the end of third grade, she was in the 44th percentile.

One night, Avital was caught late at night reading a book under the covers with a flashlight. The giveaway was a faint beam of light shining from beneath her bedroom door. "Her father and I read her the riot act," Annette told me, "but we were thrilled. The girl who could not recognize her letters was now reading on her own!"

"When Avital was five," Annette says, "her teachers had already expressed significant concern about her inability to learn the sounds of the letters of the alphabet in both English and Hebrew. The class was moving on to sounding out simple words, and she could not remember the letters in her own name. I vividly remember pointing to the letter *T* on a Tropicana orange juice container and repeatedly trying to teach her the name and the sound of the letter, to no avail."

Avital entered a first-grade Arrowsmith classroom in September 2005. At the time, she still did not know all the letters of the alphabet, in either English or Hebrew. Testing confirmed that she needed to work on all three Arrowsmith exercises related to reading.

"I decided," says Annette, "to work with her on math at home and wait for improvement in her capacity to learn before I again tried to teach her to read. In November of that year, I went to parent-teacher conferences. As is customary in Jewish day schools, I met with the Hebrew teacher first and then with the English teacher. Even with my faulty memory, six-plus years later I still remember that evening. Both teachers, separately, said to me, 'What do you think of Avital's reading?' You could have picked me off the floor, so great was my surprise. I had no idea she knew the letters of the alphabet, let alone that she was starting to read small words. I raced home that night and asked her to read to me in both languages to confirm what the teachers had said. I was not expecting her to be able to start to learn to read so quickly, and without my support."

Because Avital strengthened her capacity in the symbol recognition area of the brain, she was able to learn letters, and then quickly sight words, in both languages. Strengthened capacity in the Broca's area similarly allowed her to learn sound-symbol correspondence in both English and Hebrew. And a strengthened capacity in the motor symbol sequencing area of Avital's brain allowed her to track letters across the page, from left to right for English and from right to left for Hebrew.

She no longer needs to be retaught the material presented in class. What she learns in class, she retains. For Avital Goodman, reading has at last become like breathing.

CHAPTER FOURTEEN

NOTHING TO WRITE HOME ABOUT

I make stupid errors in mathematics and writing. I have to write things over and over again to get them right. I would work my guts out to pass in high school. I know that I've got the brains but it's hard for me to do it. I always felt that I didn't have enough time to finish my exams. I know the work but I can't get it all down if I have a time limit. I couldn't finish the aptitude test on the police exam. I froze. I had to think and write at the same time and I was dead. There is a drive inside me that I want to push forward and get ahead but I am being held back and I can't do anything about it. It is like being trapped in my head. I have all these thoughts and ideas and I can't get them down on paper. I feel smart and stupid at the same time.

GLENN SHEPHERD, *age twenty-two, 1984*

Every time I do my math instead of writing 2/6 I~~w~~ write 26. Thats called a motor problem. Its so frustuating. Its alwase in the way. When I write a storry I ~~keep making~~ make spelling mistacks or ~~I will say~~ write something I didn't plan to. Sometimes I can't speack all the words get mixed~~uu~~ up. A~~m~~t motor proplem Is ~~u~~ like being in a weel chair for the rest of ~~yu~~ your life.

MADISON PEARCE, *age twelve, grade 7, 1980*

Whaat both students are describing so powerfully is the motor symbol sequencing deficit, a neurological weakness that can impair one's ability to speak coherently, read and spell easily, and—the signature impact—write letters, words, and numbers. (The strikethroughs in the second example were in this pupil's original handwritten text. As she wrote to describe the deficit, she could not help but demonstrate it.)

To experience what the deficit feels like, try the following. On a blank piece of paper, write a lowercase letter *a*—backward. Now write the letter *a* as you would normally. How much longer did the first task take you? I would guess that it took you at least a few seconds longer. This is because you had to think about where to start, where to place your pen, what direction to move, and where to stop.

When you write the letter *a* normally, this established motor plan is encoded in your muscle memory. What allows us to write letters and numbers easily is an intricate network of cortical zones located primarily in the left hemisphere of the brain. A key zone is the premotor region. When you were in primary school, assuming this part of your brain was functioning normally, you learned a motor plan for each individual letter and later for sequences of letters, which form words. With repetition, you acquired all these plans—first for the basic task of printing letters, then for the more complex job of writing words in cursive. When you don't have an established plan for a word or letter, such as that backward *a*, writing requires conscious thought and effort.

This is what a student with the motor symbol sequencing deficit contends with every time she writes. Repetition has failed to do for her what it does for most of us: make writing as natural as walking. The mechanics of writing (forming, spacing, and sizing letters and aligning them on the page) divert the writer from the content of the writing. What is put down on the page, then, is only a truncated version of the thoughts in her head.

Luria had a poetic phrase to describe how the complex task of forming numbers and letters and eventually words on a page can happen with apparent ease. He called it "kinetic melody."

"Writing in the initial stages," he wrote in his book *The Working Brain*, "takes place through a chain of isolated motor impulses, each of which is responsible for the performance of only one element of the graphic structure; with practice, this structure of the process is radically altered and writing is converted into a single 'kinetic melody,' no longer requiring the memorizing of the visual form of each isolated letter or individual motor impulses for making every stroke." When everything is working smoothly, long-term muscle memory is created for writing each word, much as it is for riding a bicycle. Think about writing your signature: you don't need to consciously plan the forming of each letter; your name fires out automatically.

To describe the halting movements in Parkinson's patients, neurologists use the term *kinetic stutter*. The phrase might equally apply to the handwriting of individuals with a motor symbol sequencing problem. One need only look at the writing of an individual with this problem to see that the movement is halting and laborious. People with this deficit often resort to printing, which calls on simpler motor plans than cursive writing does. The handwriting sample shown here is from a forty-year-old professional writing out the Gettysburg Address.

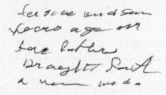

Figure 5.
"Four score and seven years ago our fathers brought forth . . . a nation conceived . . ."

In his war-wounded patients, Luria observed what happens to writing ability when the premotor area of the brain is damaged: "Handwriting begins to change; the letters forming whole words begin to be written separately, and subsequently, every stroke . . . requires a separate effort of will. . . . The order of the elements is lost and the smooth transition from one component of a word to another and the retention of the required sequence are impossible."

My name for this deficit is *motor symbol sequencing*. Each word carries important freight. *Motor* implies movement (the Latin verb *movere* means to move), and when the premotor area is working well, brain and hand work well together. *Symbol* refers to letters or numbers, the basic building blocks of math and literature that we learn as children. *Sequencing*, from the Latin verb *sequi* (to follow), suggests order and the notion of pattern—an ordered series of elements (motor impulses organized into a sequence) to produce a sequence of symbols (words).

The premotor region is involved in learning and executing the motor plans necessary for eye tracking in reading, for speech, and for writing. If you have this deficit, you cannot read quickly. Visually tracking symbols in sequence (letters or numbers on the page) requires good eye motor control, which you lack. As a result, you misread constantly—*calm* for *clam*, for example. Asked to add numbers, you misread the plus (+) sign, take it for a minus (-) sign, and of course your calculation is wrong.

In the early grades, one typical classroom assignment is to take a word from a list in one column on the page and insert it into a sentence with blanks in the other column. A pupil with a motor symbol sequencing deficit may know each word if you say it to him, but his eye-tracking difficulty makes picking one out from the list, tracking to the appropriate blank space in the sentence, and then writing it down exceedingly hard.

For these students, performing any task that involves using this area takes extra effort. The additional load on the brain means they tire quickly and can't sustain attention, making it even more difficult to track words on a page or to regain focus once lost. These children often get labeled as having attention problems, but attention is not the real issue. Lack of focus is a casualty of a weak cognitive area and the load it places on the overall system.

At a severe level of difficulty, their speech lags far behind thought, with the result that it is rambling and disjointed, wandering in the same way that their handwriting does. They could have all the information in their head to tell their story properly, but they leave out critical bits (though they think they've stated them), making it difficult for others to follow. They may stumble or hesitate or become shy and withdrawn because of this inability to express themselves.

And as we have already seen, they have trouble writing neatly. They misjudge the space they have left at the end of a line, so they do one of two things: compress those last few words into an impossibly small space, as if they were cars colliding, or break the word up, with half on one line and half on the next. They repeat words in a sentence (*the the*) or individual letters (write *weeell* for *well*). Luria explains this as losing "the ability to arrest one of the steps of the movement and to make a transition from one step to the next."

Their spelling of one word may vary, even on the same page. Spelling is affected by a motor symbol sequencing deficit in a specific way. Some students spell poorly in their day-to-day writing yet perform well on a spelling test. How can that be? During a test, the student can draw on strength in the symbol recognition area by retrieving the visual form of the word and reproducing it on the test. In daily writing, though, students need to rely more on the motor plan for the spelling of words. Stopping to recall the visual form of each word to reproduce it would slow the writing process considerably.

With this deficit, taking notes in class is problematic because students cannot write quickly enough to keep up with the lecturer. In math, when asked to divide 15 by 3, they think 5 in their head but write down 3. Their handwriting is indecipherable, and the number 4 over the course of a long calculation can morph into a 9. They write painfully slowly, so copying from the board or completing a timed test is next to impossible. They know the material, but their grades fail to reflect that fact. Measured instead is their plodding and inefficient writing process.

Because of all these errors (which are not conceptual errors but mechanical ones), they may be labeled "careless" or thought unintelligent. Glenn Shepherd described how awful it is to be "smart and stupid at the same time."

The motor symbol sequencing deficit is the most common learning disability we see at Arrowsmith. Cultural inventions such as writing, reading, and numerical representation are human-made systems introduced into the world that required the brain to adapt. Brain areas that had evolved to perform other functions were recruited to perform these new processes. In terms of writing, the premotor region in the left hemisphere, given its role in the conversion of individual motor impulses into a smoothly and consecutively organized skilled movement, was enlisted to learn the skilled motor movements (the motor plan) required to write. Writing as a process to represent spoken language developed a little over five thousand years ago, which in terms of adaptation is a relatively short time. I argue that any area that is relatively new in recruitment to perform a function is more vulnerable to malfunctioning.

The Arrowsmith exercise to address a motor symbol sequencing problem involves learning and producing motor plans using sequences of symbols. When students do this enough, the sequence will begin firing out the end of the pen without thought or hesitation. Over time, the plan becomes encoded in muscle memory, and when the student masters that plan, a new one is presented. The plans become more complex and are learned faster and faster, indicating an improvement in the capacity of the brain to learn motor plans. And what we see is this: writing flows on the page, the eyes glide smoothly over lines of text, and speech is fluent.

This class exercise involves wearing an eye patch (blue on the outside, black on the inside, though students can also customize their patch—with some opting for a leopard pattern). Students embarked on this exercise look like studious little pirates. The patch is worn over the left eye, forcing the right eye to do all the work.

Why is that important? Remember that the left hemisphere of the brain controls movement on the right side of the body (and the right hemisphere controls the left side). In the case of this deficit, we want to increase stimulation to the left premotor region of the brain. By using the right eye (and its motor movements) in coordination with the dominant hand, we increase stimulation to this left-hemisphere motor area.

The wearing of the eye patch has created some confusion, with some observers mistakenly thinking that we are training the visual field in the primary visual cortex. In the visual field, where one-half of each eye goes to each hemisphere, it would not make sense to use an eye patch since vision from each eye goes to both hemispheres.

Glenn and Madison spent many many hours wearing an eye patch and, pen in hand, assiduously tracing shapes such as Chinese characters—first simple, then more complex, then tangled and intricate. Getting it right mattered, but so did getting it done quickly. The goal was to make Glenn and Madison's writing easy and automatic.

"The tracing exercise," says Ian Taylor-Wright, who teaches English at the Toronto Arrowsmith School, "has always seemed to me one of the more meditative of the cognitive exercises. Akin to walking a tightrope, it's all about control and finesse. One develops patience and discipline through practicing it."

I remember my excitement in 1978 when I got further confirmation that what was changing was the brain and not just peripheral or surface symptoms. I was working with a nineteen-year-old woman who had a deficit so severe that the motor plans required to reproduce the English alphabet were beyond her. To enable her to work on the motor symbol sequencing exercise, I created for her a simplified set of symbols, which I named after her: the Dawes alphabet, I called it.

From reading Luria, I understood that the area being targeted by my exercise—the premotor cortex in the left hemisphere—was involved in converting impulses into organized sequences of movements necessary for speaking, reading, and writing. The exercise I created was precisely as described above, involving only the hand and one eye. But here is what I found so remarkable: as this young woman progressed through the exercise, her handwriting and eye tracking in reading improved, *and so did her speech*. No longer was her speech choppy and hesitant. No longer did she leave out large chunks of information, making it impossible to follow what she was saying. No longer did she ramble, starting one thought and not finishing it before the next thought fired out.

Prior to treatment, she took so long to express herself in speech that even friends walked away before she could finish. As a result, she had withdrawn socially. Over the course of several months, as she worked diligently on the exercise, the part of her brain that produces organized motor sequences gradually became stronger, and her speech changed. Her thoughts flowed from her brain and out of her mouth in coherent sequences, with no hesitations and no content missing. After treatment, her social life blossomed and she began dating, with the repartee flowing as one would expect from a person her age. And at no point was speech expression part of the exercise.

This, for me, was further evidence that the brain is capable of changing through the application of targeted stimulation.

I've come to meet with parents, students, teachers, and the associate head of school, Rabbi Eliyahu Teitz, at the Jewish Educational Center (JEC) in Elizabeth, New Jersey. Inside the school, its corridors flanked by tall yellow lockers on both sides, I make my way to a classroom where Rose Kandl awaits me. She has kind eyes and a tart wit. But what distinguishes her even more is the fact that she wears, figuratively at least, two hats. She is both the mother of a once-learning-disabled child and an Arrowsmith teacher at the school.

Six years ago, Rose was driving to school when she heard Annette Goodman, chief education officer of the Arrowsmith Program, being interviewed on WFMU-FM, a radio station centered in Jersey City. Rose had never heard of neuroplasticity, and she was intrigued. Her son, Josh, was struggling in elementary school.

Josh had dysgraphia, which Rose defined as "a disconnect between your brain and your hand—and the wiring is off. Learning specialists said, 'He'll never write. Just teach him how to sign his name, so if he has to sign a check, he'll be able to sign a check.'"

Neither could Josh manage a keyboard, because his coordination was so poor. He also had another label: gifted/learning-disabled. Translation: high IQ/low performance. Given a multiple-choice test in class, he would score 85 and be the first one done. But with anything involving writing, he was the last one to hand in his paper, and he would never finish. He was so slow copying from the blackboard that the teacher would have erased the board and moved on to something else before he could finish. Until seventh grade, he did all math calculations in his head and simply wrote down the answers. When he was asked in the later grades to show his calculations on paper, he couldn't do it, and he'd fail tests.

The impact of these and other learning deficits was near catastrophic. Rose paints the picture: "We switched schools—three times. In grade seven, he had a complete shutdown. He hated school. We were seriously worried about him. He was a kid who would come home and sit in his room and do nothing—with the lights out. You don't do that at twelve, thirteen years of age. You should be outside, running with your friends."

In spring 2008, Rabbi Teitz approached Rose and asked her if she was interested in teaching the Arrowsmith class at the JEC. Still intrigued by what Annette had said on the radio about the brain as an organ that can be exercised, Rose went to the American Christian School close by, in Succasunna, New Jersey, where the principal, Carol Midkiff, has had the Arrowsmith Program since 2007.

"I saw the program," Rose told me, "and I wasn't quite sure what I was seeing, to be honest. I remember sitting there looking at a pupil doing the clocks exercise. I said to myself, 'I should be able to read this clock, but it has too many hands.' I couldn't figure out what it was that I was supposed to be doing. I spoke to Dr. Midkiff. I spoke to the teachers. I remember vividly there was a second-grader in the program, and her teacher told me, 'This girl was in the bottom of the class in math, and now she gets straight As.' And all the teachers were saying how wonderful the program was. Dr. Midkiff was telling me how wonderful a program it was. And I came back and I said to Rabbi Teitz, 'Okay. If it works, this is the best thing since sliced bread.'"

It was decided that the JEC would implement the program and that Rose would go to Toronto that summer to become a certified Arrowsmith teacher. I asked Rose to recall that experience and had to laugh at her reply—which was surely hyperbolic to make a point. "I cried through the whole thing," she said. "It was intense."

But taking the Toronto course had two other effects. It redoubled her compassion for children who struggle with learning disabilities, and it made her absolutely certain that Josh had neurological deficits and that he could be helped by this program.

"I was sitting in the training," Rose remembers, "and I'm listening to all these different areas of dysfunction. I'm saying, 'That's my son Josh, that's my son Josh.' All of these difficulties and nothing, except for his reading, had been addressed in all those years. I was sitting there saying, 'This is amazing. This is exactly what I've been saying about my son, but nobody's been listening to me.'"

Our own testing indicated that Josh did indeed have a very severe motor symbol sequencing problem, which explained the poor handwriting (it began to improve three months into the program). By the end of the year, Josh had made enormous strides.

Finally, Josh was retested by the school board and declassified as learning disabled. No more compensations. No more need for extra time.

"Arrowsmith," says Rose, "changed Josh's life. This was a child whom we had to literally drag out of bed to get up in the morning. He never wanted to go to school. He never wanted to interact. Everything was so difficult for him."

Now in tenth grade, Josh writes papers independently and gets As and Bs and even A-pluses on his English essays. He can't wait to go to school and he talks at home about his teachers and friends there.

"He's just a completely different child," Rose says. "I like him now. I loved him before, but now I like him, too."

The motor symbol sequencing deficit is painful in every sense of that word. One of our students used to tell his mother reassuringly that his woeful handwriting was no big deal. "Doctors get away with it," he would joke. "Why can't I?"

But it's no laughing matter. The inability to write can become a source of acute embarrassment and mental anguish. And because the brain-muscle link is so compromised, the act of writing may also be physically painful. One student put it this way: "There's an avalanche behind my hand and I'm trying to control it."

Stuart Davies (the accountant we met in Chapter 9) kept a handwritten journal when he was a student at Arrowsmith some seven years ago. Here's his entry from May 20, 2004:

I am so excited to write about my improvements. First, look at my writing. The "chicken scratch" is gone. Wow!!! My writing is definitely neater, but more importantly, the writing process is not painful. I no longer have to run to a computer keyboard to get around my messy writing and my writing pain. The flow from my brain to my pen is much smoother because my mental "gridlock" is gone. I used to think that my brain was ahead of my hand, but I am happy to say when I write now, my brain and my hand are working together nicely.

Some of the changes that follow Arrowsmith exercises may appear gradually, but they can also strike like a bolt of lightning. When Ann Tulloch and Devorah Garland worked on the motor symbol sequencing exercise, they underwent cognitive change that felt like an awakening of the aesthetic senses.

Ann's severe motor symbol sequencing challenge meant her right eye didn't work well, which in turn meant that her two eyes didn't work together in order to mark depth. As a result, everything for her was one-dimensional.

"It wasn't that I didn't see things," Ann told me. "It's so hard to explain. I guess it was like a gray; it was flat."

Here's how Ann Tulloch described her transformation: "I was on the streetcar, and all of a sudden it was like I was in a 3D movie. Everything just . . . all the colors, all the shapes and my sense of distance—everything jumped out at me. Everything was super-bright; shapes were really pronounced. All of a sudden, I could see all the details. I was so happy. I could see that buildings had sides and backs; there was depth. Before it was like looking at a movie set; buildings just had fronts. There was no depth, no perspective."

Ann was so unnerved by the change that she called me to ask if she was going crazy. I assured her that as far as I knew, her sanity was not in question. She simply now had what I call dynamic vision and a new way of perceiving the world.

Now that Ann's eyes were working together, she could use binocular cues, requiring input from both eyes for depth perception. The brain integrates the two-dimensional images from each eye to create a sense of three dimensions (a process called *stereopsis*), which allows people to perceive depth. Any disorder in which the eyes do not align properly can also interfere with depth perception. In cases of severe motor symbol sequencing, we often note that the person has a less dynamic right eye. This is because the left premotor region is involved in controlling the motor movements of this eye. Ann likened the change to going from hearing monophonic to stereophonic music. And to remember what it was like before, I told Ann, "All you have to do is close your right eye."

"I feel like there's so much more," Ann told me, "and I feel like I want to make up for lost time. I always appreciated things like art. But there was always a block there, something I missed, and I never knew what it was. I just couldn't really appreciate a good piece of art or a good piece of furniture. It's like I've always walked around with all these walls around me, and I could never get out. Now I feel like they're starting to come down. I'm getting a taste of it, and I want more and more and more."

At the three-month mark of her own cognitive program, Devorah had the same kind of epiphany. It shocked her. "I was walking down Yorkville Avenue in Toronto," she told me, "and everything looked different. I was looking around at the buildings, and all the shapes were different. I just walked along, and everyone else was saying 'Hurry up, hurry up.' And I'm saying, 'Can't you see it?' My perception of the world had become sharper. Shapes, textures, and colors were clearer and had a luster to them I had not noticed before. I remember walking down the street staring at everything, feeling such wonder."

The brain exercises that Ann and Devorah had been doing had changed the functioning of each woman's brain, and one discernible impact, among others, was that they now experienced stereoscopic vision and depth perception where before there was none.

The process can sound instantaneous, but it's not. Underlying change in the individual's cognitive function takes place gradually and often without that person being aware of it. The changes need to reach a critical point before the person experiences what Ann and Devorah experienced: a radical shift, in their case, of depth perception.

When handwriting goes from ragged and disorganized to neat and fluid, it doesn't take a handwriting expert to know that something has changed. This change is in the brain. And noticeable change begins to occur within a few months of commencing the cognitive exercises. As students work on the motor symbol sequencing exercise and functioning in this area improves, we often see a spontaneous transition from printing to cursive writing. The writing samples on the next page illustrate the change.

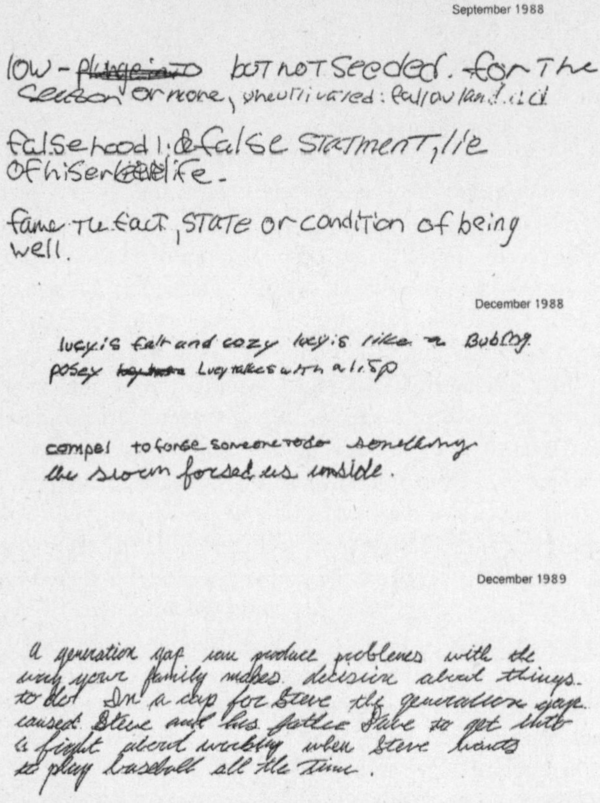

September 1988

low – ~~flagginto~~ but not seeded. for the
season or none, ynwotivated: Pallow land. etc.

falsehood 1. & false statment, lie
of his or her life.

fame the fact, state or condition of being
well.

December 1988

Lucy is felt and cozy lucy is like a Bubloy.
posey ~~together~~ Lucy makes with a lisp

compel to force someone to do something.
the storm forced us inside.

December 1989

a generation gap can produce problems with the
way your family makes decision about things.
to do In a day for Steve the generation gap
caused Steve and his father Gabe to get into
a fight about wanting when Steve wants
to play baseball all the time.

Figure 6.

Within three months of starting the exercise, this fifteen-year-old student's letters became more uniform in size. In the second sample (from December 1988), most of the letters are appropriately small—in sharp contrast to the September sample, where the letters alternated between large and small. Spacing between words had also become more consistent, and the student is beginning to alternate between printing and cursive. By December 1989, the student no longer prints, and his cursive writing is uniform and legible.

One seven-year-old girl had a motor symbol sequencing deficit that was so severe she would stay up at night doing her homework over and over again, trying to make her writing look neat. Her mother found her in the bedroom closet with the light on and about a hundred crumpled pieces of paper all about her.

This girl started the cognitive exercise to address her writing problem in September 2005, and by December she was reporting that her hand hurt less when she wrote, and soon after, the pain disappeared entirely. The Arrowsmith file on this girl contains this note for January 17, 2006: "Mother reports that her artwork has dramatically improved, there are more details, it is smooth and flowing, and it is above average instead of juvenile. There is a huge leap in her skill."

Naturally the girl uses the same hand for drawing as for writing, and as she developed more control in that hand, this improvement was seen in her art. In the middle of seventh grade, her teachers recommended that she skip eighth grade and go straight to high school, which she did. Her English teacher, who was particularly fulsome in her praise, said that this student's writing style and structure were as good as any eighth grader's work she had ever seen.

Over and over, the testimony is the same: writing begins to flow, the intended thought is translated into the motor symbol sequence, and the words appear on the page. One mother said that her daughter's writing was labored, even when she wrote her own name. Now, she says, her daughter's hand "flies over the page."

CHAPTER FIFTEEN

BLIND TO ONE'S OWN BODY

As I continued to work with learning-disabled children, I gained a more refined sense of the connection between a certain kind of learning challenge and a particular function of the brain.

In 1978, I looked back, for example, at my lifetime of clumsiness and klutziness. I collected incidents and catalogued and analyzed them, while becoming more convinced that they were all linked. Eventually I would give my problem a name: *kinesthetic perception*.

The entire left side of my body was my nemesis. I could not wink with my left eye. On my left side (and just my left side), my teeth often cut into my tongue and that side of my mouth without my noticing—until I ate salty food and was jolted into a stinging realization.

I was constantly over- or underreaching with my left hand. Picking up a pen, I could not at times stop the movement in its arc once begun and would end up throwing the pen over my shoulder. I once leaned over in a lecture hall to pick up my pen and my body kept going until I fell onto the floor. Trying to tuck my hair behind my left ear, I would accidentally scratch my face or eye with my left hand, so I kept my nails short to prevent self-inflicted injury. For that reason, I never used that hand unless forced to. It would do things of its own volition.

We were taught to touch-type in high school, which meant we had to ignore the keys and focus on a visual board at the front of the class. My left hand was a kind of Mr. Hyde to my right hand's Dr. Jekyll. The fingers on my left hand constantly crossed paths. I could not judge how far to reach with that hand in order to hit various keys, and I invariably hit the wrong one. I was amazed that people could actually type without looking at their fingers. My teacher would come by, lift my head up, and point it in the direction of the keyboard chart on the wall at the front of the class. I miserably failed that class, earning a grade lower than my age.

Even eating revealed my lack of control over that left hand. My father, a kind man who cherished his only daughter, could not help himself. "Don't eat like a Philistine," he would say. I would invariably use too much pressure when holding the fork with my left hand while cutting with the knife in my right, causing me to loudly scrape the plate or inadvertently knock food off the plate. I could not reliably hold a cup or glass in my left hand; I dropped it more often than not. On many occasions I would spill the contents, since I could not judge when to tip the container relative to my mouth. I frequently banged my teeth with a cup or glass.

I avoided sharp implements, for I had cut my left hand with a paring knife too many times. When I was eleven, I misjudged how close my left hand was to an electric mixer and got my finger caught in the moving blades. That hand just seemed to get in harm's way.

In my mid-twenties, I was once getting into my car and relied on my left hand to close the door. But the physics was all wrong: I wasn't yet installed in the driver's seat when that left hand of mine closed the car door on my head, thereby shattering the crown on my front tooth—a crown that had been placed there after an incident at a swimming pool when I was sixteen. I was on the low diving board and lost my nerve. As I was descending the ladder, I misjudged and put my left foot between two of the steps rather than on the step, lost my balance, and landed face first on the concrete floor, shattering that unlucky front tooth.

I had ample evidence that the relationship between my body and the world, and even parts of my body with each other, was a rocky one.

The left side of my body was so alien that I could feel pain on that side and have no inkling of its source. I once put my left hand on a hot stovetop burner but was unable to react quickly. I did the same thing with my left foot on a hot radiator. I learned the hard way that when I experienced pain, I'd have to visually search for its source. The left side of my body was a patchwork of bruises, and I had no awareness of the incidents that had caused them.

Nor was I very good at sports. Another neurological deficit, one I would come to label "spatial," joined forces with my kinesthetic deficit to erase any hope of athleticism. That little three-year-old girl who cut her head open on the family sedan in the driveway was ill prepared later on to hit or catch a baseball. Gym was a nightmare. Running at the pommel horse, I was never certain if I would clear it or crash into it. Anything that required coordinated body movements, such as square dancing, was beyond me.

I would frequently trip over my left foot when walking. My classmates called me a klutz, and I was always picked last for any team sports.

In only one sport did I excel: badminton. I discovered it in ninth grade, and what I liked about it was the fact that it seemed a slow, gentle sport. Badminton gave me what I always needed more of: time—time to compensate. Swimming was the other sport I was good at, and I became a lifeguard. Swimming did not require quick or precise movements, and if I chose times when the pool was not crowded, I avoided the potential problem of colliding with others.

When waterskiing one time, I misjudged the proximity of the dock and crashed into it—left side first. I had many falls to the left off bikes, and I remember a particular crash while cycling down a steep hill near our house. On another occasion, when I was thirteen, I came flying down a hill on a skateboard and applied too much pressure on my left side, causing the board to dip and strike the curb. In the ensuing fall, I fractured a finger and severely bruised my body.

Nor could I walk with anything approaching grace. I could not judge the amount of pressure I was applying as my feet hit the ground. My father said I walked like an elephant. High heels were never an option for me.

And I was terrified of heights. When we took trips to visit my parents' families in British Columbia, we would sometimes stop the car and everyone else would get out to admire the view from a mountain pass in the Rockies while I cowered in the car. If my brothers coaxed me out, I would keep twenty feet back from the edge. (Later, when I learned to drive, my knuckles on the steering wheel would be white if I had to navigate a narrow mountain road with a sheer drop on one side of the car.)

Even left and right confused me—for three reasons, all of them cognitive: I had no awareness of sensation and where it was arising on my body (owing to my kinesthetic problem); due to my spatial deficit, I had no spatial map of my body; and due to my symbol relations deficit, I had no concept of left and right.

There was also the humiliation of taking forever to learn to tie my shoelaces. When I finally mastered it, I was eight years old (five to six is the norm), and it had taken an enormous effort and an endless cycle of learning this small skill before it finally stuck.

As with my other neurological deficits, I became my own guinea pig. Thanks to Luria, I was able to understand which part of my brain governed my kinesthetic awareness (or, in my case, my lack of it), and I developed an exercise that I hoped would stimulate that area of my brain.

If what was not working in my case was the part of my brain that provided sensory feedback to help me learn how to move my body with precision, then I needed to find a way to work that function. I knew that I used my vision to compensate for not knowing where my limbs were at any one time, or my entire body for that matter. So I knew I needed to create an exercise that relied on sensory feedback without the support of sight. I thus developed an exercise that involved performing precise movements with my eyes closed. I knew I had mastered the movement when I could perform it as precisely and accurately with my eyes closed as when my eyes were open. I then made the movements more complicated and did them ever more quickly.

Rigorous and repeated exercise paid off. Gradually I began to feel subtle variations in movement. My brain was informing my muscles about how far and where they needed to move to perform the skilled movement.

Now I'm very comfortable using my left hand. I can pick up things, I can wash dishes without chipping or breaking them, I can drink coffee without fear of spilling, and I can use a fork in that hand, which is much more coordinated than it was. I now type at eighty words a minute with very good accuracy.

The word *left* in Latin is *sinister,* a word we now use in English. But the word has come to mean that something harmful is either happening or about to happen. For almost thirty years of my life, I was imperiled by the left side of my own body. The rewiring of my brain's right hemisphere by repeated and specific exercise restored my left side and brought it into harmony with my right.

The two sides of my body get along well now.

If you have a kinesthetic perception deficit, your brain is not registering or interpreting sensory information from your own body. One result of this deficit is that it's difficult for you to know where one part of your body is in relation to the other parts. For example, if the left side of your body had been significantly impaired, you wouldn't register the angle of your left arm or how far that arm extends from your body. And even if the right side of your body was functioning normally and you were correctly registering sensory information from your right arm, your right and left arm could not work together. Imagine trying to learn a complicated dance step when the left side of your body is taking ten times longer than your right side to learn the positions.

With the kinesthetic deficit, sometimes only one side of the body is affected, sometimes both. Sometimes just one hemisphere of the brain is involved, sometimes both. People with this deficit are often called "klutzy" and "clumsy," but those words fail to capture the depths of the disorder or the way it puts you at odds with your own body.

I remember a mother telling me about her young son who was very excited when she bought him an ice cream cone. But when he brought the cone to his mouth, he missed, hitting the side of his face with it, and the ice cream fell to the ground. The distraught boy was later found to have a severe weakness in kinesthetic perception.

In *The Man Who Mistook His Wife for a Hat,* Oliver Sacks describes a woman who had lost her "position sense." In this case, the woman's parietal lobes (more on this later) were working fine, but owing to nerve damage, the sensory information wasn't getting to the brain. *Proprioception* (a word that derives from the Latin *propius,* meaning "one's own," and another word, *perception*) is the word that Sacks had used to explain to the young woman the disconnect between her body and her brain. I found her response poignant and familiar. Here is what she told Sacks: "I've already noticed that I may 'lose' my arms. I think they're one place, and I find they're another. This 'proprioception' is like the eyes of the body, the way the body sees itself. And if it goes, as it's gone with me, it's like the body's blind. My body can't 'see' itself if it's lost its eyes, right? So I have to watch it—be its eyes. Right?"

Loss of proprioception at the muscle, tendon, and joints, not the brain, was the problem here. Cut off due to nerve damage, information was unable to reach the brain. In the cases described here, the problem is the reverse: the brain is unable to register and interpret sensory information properly. In each case, the person's experience is the same—being blind to one's own body.

The somatosensory area of the brain also discerns the shape of objects being held in one's hand and distinguishes subtle differences between objects that touch the skin. For example, when reaching into my pocket with my left hand, I was unable to identify an object, even my own keys. Think of the game that children play, drawing a letter, number, or shape on someone's hand that the recipient must identify from touch. This would be impossible for someone with a kinesthetic deficit. And imagine learning to read braille; the raised patterns of dots require tactile perception to discriminate the letters.

To get a feel for this deficit, try the following simple experiment. Put your coffee cup on the table within arm's reach. With your eyes open and your upper body still, touch your nose with your index finger and then reach out to touch the top of the cup handle. Now, without moving the rest of your body, bring your arm back to your body, touch your nose, close your eyes, and again reach for the same position on the cup. Can you find it? Do you over- or underreach, or are you on target? Try it with either hand. Is there a difference? Then try the same thing with a coin or a smaller target. If you have no difficulty with this maneuver with your eyes closed, your kinesthetic perception may well be fine; if you struggled with the experiment, it's a possible sign of kinesthetic weakness.

In simple terms, what I'm describing is the ability to feel, and remember, the sense of movement. Receiving and processing sensory input or feedback occurs at the most basic level, such as knowing where sensations (pressure, touch, temperature, or pain) are occurring on the body. A more complex process is involved in knowing where your limbs are in relation to your body and how your body is positioned in space.

As you move through space, continuous feedback provided by sensory input about the position of your body helps guide your movements. And if the kinesthetic area is functioning as it should, current movements can be guided by the memory of sensory information from past movements.

To understand how this works, think about learning to type. At first, the fingers and hand need to learn the position on the keyboard, the amount of pressure to apply, the distance the fingers need to move to reach the different letters. After much practice over time, a sensory memory for the feel of the movements develops and allows the typist to be proficient.

Sensory feedback is critical in learning a motor skill (the backhand in tennis, for example), and without that continuous loop, you never learn efficiency or accuracy of movement. The feedback can be likened to a teacher at your side, helping you to adjust and modify until, with enough practice, the sequence of muscle movements can be repeated with ease. The sensory memory of past movements becomes our guide.

Writing is another motor skill, so it too depends on kinesthetic input. In the act of writing, an internal sensitivity to the movement required to write a letter reduces the child's need to visually monitor her fingers or the point of her pencil while writing. Similarly, a pitcher on the mound about to throw a ball may appear to be acting instinctively and automatically, but in fact there's a lot going on cognitively. The parietal lobes, the part of the brain that governs spatial and kinesthetic perception, are called into play. Almost without thinking, the pitcher is gathering information on the position of his own body in space and how it relates to the positions of the catcher and the batter. Sensory memory, meanwhile, is calling up precisely what he did the last time he threw that pitch: the tightness of his grip, the grip itself, the moment of release. All of this information is necessary for that ball to go where the pitcher wants it to go.

The somatosensory cortex, located on each side of the brain (the front part of the parietal area in the midline of the brain, running from the top of the head to each ear), receives sensory input from the opposite side of the body. Thus, a problem in the right hemisphere of this cortex will show up as a sensory deficit on the left side of the body, and vice versa.

Within this cortex is the "sensory homunculus" (or little human), a cortical map that directly corresponds to different parts of the body and processes the sensations they send. If a problem is localized in a specific area of this cortex, it will affect sensory perception only in that corresponding part of the body.

Location of the somatosensory cortex

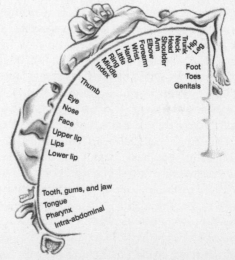

Figure 7. Sensory Homunculus.
This map shows the location and amount of space in the brain allocated to processing tactile signals from different parts of the body. More sensitive areas, such as the lips, have more cortical real estate.

Small but sensitive areas of the body such as the lips, tongue, thumb, and fingers take up more cortical real estate than do less sensitive but larger areas such as the trunk. This makes sense if we think of the importance of these sensitive areas in speech and toolmaking.

Along with the neurological deficit in the area governing kinesthetic perception on the left side of my body, the area of the somatosensory cortex that registers sensory information from my lips, tongue, and mouth was not working properly. I called this a *kinesthetic speech disorder* to note this specific area of impact. Not all people identified with a kinesthetic perception problem have speech involvement, as it depends on which part of the somatosensory cortex is involved. The result is that oral motor movements become diffuse and imprecise, leading to blurred speech articulation. People would say I slurred my speech, especially when I was tired. I had to concentrate when speaking in order to enunciate my words. As with my other neurological deficits, I developed an exercise, one involving speaking word combinations that tied my tongue into knots, to work that weak part of my brain. I don't slur my words anymore.

To experience what it feels like to have a speech kinesthetic disorder, place a few marbles in your mouth (cherry or olive pits work equally well) and repeat, "She sells seashells by the seashore." Note how much attention you must pay to clearly enunciate each word when your mouth and tongue must work around these objects. Patterns of sensory feedback previously learned for making these sounds no longer apply. You must pay careful attention to produce each sound; many you will utter imprecisely. For someone with a speech kinesthetic problem, an active effort of will is required to prevent one sound running into and over another.

Cicero proclaimed the ancient Athenian statesman Demosthenes "the perfect orator." But in his youth, Demosthenes had a speech impediment. His speech was described as inarticulate, and he was not easily understood. To overcome this problem, he put pebbles in his mouth and then forced himself to speak clearly despite the pebbles' interference. It is impossible to know if his problem was what I call speech kinesthetic, but the features are consistent with this disorder. Demosthenes created his own exercise, forcing attention to the sensory feedback of the muscles of the mouth and tongue. It worked: he had a long and successful career as an orator.

Someone with a speech kinesthetic problem would have difficulty in finding the mouth's articulatory positions in order to pronounce different phonemes. The primary confusion is between phonemes that are similar in articulation, such as *th* and *f*—so the word *thirty* would be said as *firty*. If the brain map for lips or tongue is severely compromised, individuals drool as if they had been to the dentist and had part of their mouths frozen. This has been described by some clinicians as "positional apraxia," though Luria called it "afferent motor aphasia."

Andrea Peirson, a teacher at the Toronto Arrowsmith School, tells the story of a young woman named Elinor Curtis who came to the school in 2003. Says Andrea, "She told me one day that she was always a klutz, she was always falling over desks, always knocking things down, always banging into people and things and dropping things."

As I had done in my youth, Elinor played badminton—and perhaps for the same reason. The birdie drifts down slowly and doesn't come at you with the same velocity as a ball or puck, allowing extra time for the body to get into position to hit it. Remember that if there is no sensory memory associated with particular movements, much trial and error is needed. Though she was not a skilled player, Elinor found that badminton offered her release from a punishing lifestyle, one that was completely built around trying to compensate for several major learning difficulties. Here's how Andrea described that life: "Elinor was spending upwards of forty hours a week on top of school just to study for tests. She had writing difficulties. She couldn't remember what the teacher had been talking about in class. Everything she learned she had to relearn three or four times. She was seventeen and burnt out, completely exhausted. The only thing she did was badminton on the weekends, and that was it; she had no life. And she was getting marks in the 90s."

Elinor came to Arrowsmith and over the course of two years addressed her many learning dysfunctions, including the kinesthetic perception problem. The delicate tracing exercise we use at Arrowsmith to address this deficit at first posed a huge challenge for her. Elinor pressed too hard or too lightly, and she found it extremely difficult to master intricate movements of the hand with her eyes closed. But over time, the exercise led to improvement in all her physical activities. Where Elinor's movement had been broad and unrefined, it was now precise and skilled.

This is what I believe is happening in the brain of someone doing this cognitive exercise: through repeated sensory feedback, the cortical map in the somatosensory cortex becomes more precise. If, at a simplistic level, we think of the problem as resulting from a poorly charted map giving overly general and at times inaccurate sensory information, this exercise reworks the map, giving it precise coordinates. As the map becomes more refined, the feedback becomes progressively more attuned to subtle differences in movement that further enhance the map.

Elinor struggled with the kinesthetic exercise in the beginning, but then she got better—and so did her badminton game. Her badminton teacher had been asking her to do particular physical exercises, and there were some things that she could not do. She could never make her body move in particular ways, for example—but then she started being able to do it. She had more range of motion and a better swing because she could now refine her movement based on immediate sensory feedback. And as she continued to practice, her new movements were guided by the memory of sensory information from previous movements.

At the end of the year, she won a badminton championship. And her writing changed too; she no longer pressed so hard with the pen on the paper. Her brother did this also; his kinesthetic perception deficit was more severe than hers. When he wrote with a pencil, it would break, and when he used a pen, the ink line was very thick. If he had stacks of paper, you could see the imprint of what he was writing through three or four pages.

The tracing exercise rewards delicacy of touch, but part of the exercise has to do with remembering the feel of the initial movement with eyes open. Elinor would sing or count or find other ways to help her remember. With her eyes open, she would, for example, begin to trace and at the same time start singing a particular song and match the moment she stopped tracing with a certain word in that song. Asked to repeat the tracing with her eyes closed, she had her cue for when to stop. This was ingenious, of course, and to be expected from a very intelligent young woman who happens to be a perfectionist and had been compensating all her life. But her singing was also doing an end run around the exercise I had devised to stimulate the part of the brain governing kinesthetic perception.

Andrea had to tell her, "You're compensating, Elinor. That's not going to benefit you."

Andrea was using the word *compensating* in a particular way—one that bears a little scrutiny. Throughout this book, I have used the word in its traditional sense: to make up for, to offset. We all have cognitive weaknesses and cognitive strengths and use the latter to counterbalance the former.

What Andrea meant in this case was that Elinor was drawing on a strength in cognitive areas related to musical timing to know when to stop tracing a line. She accomplished the trace, but she was missing the point of the exercise: stimulating the part of her brain that governs kinesthetic perception without support from other cognitive areas. So Elinor would do something else, and later Andrea would come to me with the question: "Is this compensatory or not?" And Elinor would routinely want to know, "Am I compensating? Am I compensating?"

That, ultimately, is how the Arrowsmith exercises and methodology manuals came to be developed in the late 1970s and early 1980s. They were built and refined, case by case, deficit by deficit. I worked from Luria's descriptions of the nature and function of the brain areas and learned from stories such as Elinor's. That process continues.

One student with a kinesthetic perception deficit was a butcher. All day long he handled razor-sharp boning and trimming knives, but had little sense of where his hands and fingers were. When he first came in, his left hand was covered with bandages and scars. People with a kinesthetic problem will use vision, as he was doing, to compensate. Slowing down and being careful and watchful were time-consuming and, as evidenced by his scars, not entirely successful. After two years of working on this area, he was able to cut meat without cutting himself.

Let's go back to June Winters, the woman introduced earlier in the book who was bewildered by simple questions from a bank teller because of her thinking deficit. Another issue that caused her grief, in an entirely different way, was a kinesthetic perception weakness.

"June drove in Canada, but a car with an automatic transmission. When we moved to Europe, I got a car with a stick shift," her husband, Robert, told me, "not realizing that she would have difficulty with a manual shifter. And she did have great difficulty the several times she tried to learn. It's been fifteen years now since she has driven, and now she's afraid to drive."

June has a mild to moderate weakness in kinesthetic perception on both her left and right sides. She is not completely aware of how her body is positioned when she's moving through space. In going through a doorway, for example, she doesn't know exactly where both sides of her body are and often ends up colliding with a door jamb. When June heard me explain this deficit, a light seemed to go off in her head.

"Is that why the driving instructor always kept telling me to go closer to the center because I was going too close to the side and I might drive off the road?"

When you drive a car, the car becomes an extension of your body. If you are affected on the left side, for example, your car may drift to the left as you drive, and that side of the car may tell the tale with dents and scratches.

Imagine that you are approaching a desk. You get information about where you are in relation to the desk from the intricate interplay between the body's sensory receptors and the brain's processing of the information they yield. If your brain fails to process that feedback, you won't sense that your body has shifted; perhaps it's begun moving at an angle rather than in a straight line. Without that moment-by-moment sensory feedback telling you where parts of your body are in any given space, you can't self-correct, so your toe meets the desk, or worse.

It's the same with operating a clutch and a stick shift while driving a car. Here, your brain uses the sensory information of touch and pressure to tell your leg when to release the clutch. June would often ask her husband how he knew when he had the clutch at the right height. And he would tell her that he "just knew." Because his brain was properly processing sensory input from his body through repeated experience, knowledge of how far to move the clutch and the amount of pressure to apply was mapped out in his brain. Releasing the clutch at the right moment had become automatic for him.

A kinesthetic perception deficit prevents mapping this kind of physical experience and retaining that map. June also has difficulty gauging how hard she is pressing on the accelerator, and sometimes her car fishtails and hits the curb. June plays the piano but admits she fumbles the keys—and until she addresses her kinesthetic deficit, no amount of practice will fix that.

June has both kinesthetic and spatial deficits. She is also understandably and rightfully wary of cliff edges, balconies with low railings, and the perimeter of the flat roof of a building. With this combination of deficits, there is an overwhelming sense of losing control, of misjudging distances, of taking a misstep and falling. This is not a phobia, a baseless fear that can be addressed on the therapist's couch. Genuine and warranted, June's fear is a survival mechanism. She suffers from the same combination of cognitive deficits that I did—deficits that made me terrified of heights and mountaintop lookouts. Unable to gauge where her body is relative to the cliff edge due to a spatial deficit and unable to count on coordinated movement from her body owing to a kinesthetic problem, June stays clear, as I did.

Those burdened with the kinesthetic perception weakness frequently trip when sidewalks are uneven since they can't feel slight dips and rises. They feel more comfortable walking to the left or right of another person, depending on which side of their body has the deficit, and they may push people off the sidewalk since they can't walk a straight line. They wear down the letters on their keyboard and heels on their shoes because they press or step too hard. They might have receding gums because they brush their teeth too hard. A woman with a kinesthetic deficit may be told by her doctor to do a breast self-examination every month, but how is that possible when she can't feel any change in the tissue?

Walking in high heels, shuffling a deck of cards, eating with chopsticks, drinking from a water bottle, or pouring liquids from one container to another: all are difficult for those who have this deficit. Riding a motorcycle likewise poses dangers for someone with a kinesthetic deficit. One student came to us after enduring three road accidents—all in the first month he owned the motorcycle and all because the movement of the machine and the movements of his body were perilously at odds with one another.

Those with a good kinesthetic sense take for granted the things they can do: touch-type, perform the fingering to play a musical instrument without looking at it, dance a complicated step without looking at their feet.

The kinesthetically blessed often choose professions where that gift can shine. They become professional athletes, race car drivers, surgeons, magicians, dancers, or massage therapists. But at elite levels of sport, even a mild cognitive impairment can be enough to keep you off the medalists' podium.

We once helped an Olympic skier who had a mild kinesthetic deficit that left him habitually making minor misplacements of his body on the left side so that he always fell on that side. We also helped a show jumper who had a mild kinesthetic weakness, also on the left aside. Her inability to control the amount of pressure to the horse's left flank had cost her in competition.

Intimately connected with the kinesthetic area of the brain is the primary motor area, which sends nerve impulses to the muscles, telling them to move. This area directs which muscles should move and how fast, in what direction, and how much force those muscles should apply. The kinesthetic area provides feedback to tell the muscles what they have just done: how far and how fast they have moved, in which direction, and how much force they have applied, all in the interests of guiding and modulating movement.

A problem in the primary motor area interferes with the speed, strength, and control of muscle movements on one side of the body or the other, depending on which area is affected. Think back to the homunculus of the somatosensory area; there is a similar map in the primary motor cortex in each side of the brain that corresponds to particular parts of the body. And as with the somatosensory area, the control is contralateral: the primary motor area in the right hemisphere controls muscles on the left side of the body, and vice versa. People with this deficit have general muscle weakness and slow reflexes, and their movements are less coordinated than the norm. The traditional phrase for this is *low muscle tone*.

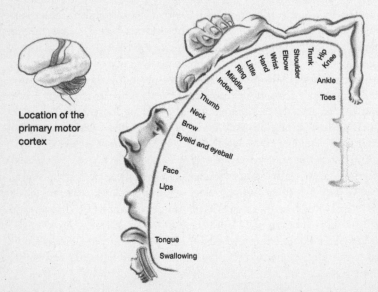

Location of the primary motor cortex

Figure 8. Motor Homunculus.
This map shows the location and amount of space in the brain allocated to controlling the movement of different parts of the body. Areas requiring more refined motor control, such as those involved in speech and hand movements, have more cortical real estate.

The story of Kirstin Lindholm illustrates how this deficit plays out. She was eighteen when I first saw her, and she moved as if she had suffered a stroke. Her primary motor deficit was so severe that it affected all movement, from walking to using tools. Her whole body was flaccid, and her movements were slow and required much conscious deliberation to carry out. When sitting in a chair, she slouched because she had almost no muscle tone. Both the right and left sides of her body were impaired. Kirstin lacked the muscle control to hold a pen properly and used her whole fist when she wrote. When she squeezed my hand, her grip was flaccid—as the saying goes, "limp as a dishrag." People watching her assumed that she was either intellectually impaired or inebriated, and they avoided contact with her.

The Arrowsmith cognitive exercise for this deficit involves fast and controlled transitions from one muscle position to another. Primary motor neurons must fire rapidly to execute movements with precision, and those movements become progressively more complex, placing more and more demand on this cognitive area. As the area improves, motor control improves.

After working on the primary motor exercise, her movement became stronger and more controlled. She tried knitting, which she had always wanted to do, and she was successful, much to her delight. Kirstin was so excited about demonstrating the increased strength in her hands that she had to be cautioned not to crush people's hands when she shook them. I thought she had fractured the small bones in my hand, so strong was her grip. Kirstin was able to join her family skiing and became very proud of her skiing prowess. There would be no more sitting in the chalet at the bottom of the hill.

CHAPTER SIXTEEN

A SCHOOL TAKES SHAPE

1980

Over time, I developed more cognitive exercises and the small number of students I was working with grew. With furniture from the Salvation Army, Joshua and I launched a school. I exhausted my savings, and we rented a 1,000-square-foot space in downtown Toronto on Yorkville Avenue. I used my middle name, Arrowsmith, for the school's name—to honor my Arrowsmith grandmother's pioneering spirit. I ran the school, while my brother Donald looked after the technical side of things as the brain exercises migrated from flash cards to computer screens. That first year, we had eight students aged twelve to eighteen.

The Arrowsmith model would take children out of the mainstream, address their learning challenges, and as quickly as possible get them back with their peers. This was the launch of what I called a cognitive classroom. A cognitive classroom serves a vastly different function from a regular classroom: the latter is where you go to learn things; the former gets the brain ready to learn. More than three decades ago, that approach was seen as radical by most, groundbreaking by some.

For my master's thesis, I had done an analysis of the outcomes for children diagnosed with either behavioral and emotional problems or learning problems. I found, on follow-up nine months to five years later, that therapeutic intervention for the children identified as having behavioral and emotional problems was successful for 66 percent of the children. But students struggling academically, those who were achieving below their age-expected grade level on academic tests administered during their initial assessment (mathematics, word recognition, spelling and reading comprehension), continued to perform poorly in the same subject areas in relation to their peers. "The amount of remedial intervention was not related to change in academic performance," I wrote then. Simply put, traditional intervention was not working.

I remember going through Luria texts with a yellow highlighter in my hand. I have many of his books that were translated into English (he wrote a score of books and hundreds of scholarly articles), and every one of those texts in my library is heavily marked and annotated.

Then theory turned to practice as I began to experiment with creating cognitive exercises: the clock exercise was first. Two things were happening at the same time: I was working on my own learning problems and with students on their learning challenges. It was as if each of us was in the gym at a different exercise station (running on the treadmill, lifting weights at the bench, doing back flips on the uneven bars). Each of us had a tailored program, and each of us had specific goals in mind.

Joshua and I had no money and a tiny clientele, and the dynamics with him were always difficult, but I look back on this time as one of the most stimulating in my life.

Then encouragement came—from several quarters.

First, a vocational counselor at the YMCA in Toronto, Reg Bundy, was seeing many young adults whom he knew to be learning disabled. He became aware of our work at Arrowsmith and started referring people to us; their siblings often became students too.

A school principal in Stouffville, a community north of Toronto, heard about us. He was seeing the emotional damage wreaked by learning disorders, and he referred students to us. Word of mouth was spreading the news of Arrowsmith.

By degrees, the Toronto school expanded. But Joshua, who had grown up in Montreal, had always dreamed of living in New York. And in 1990, an Orthodox Jewish family from New York came up to Toronto to investigate our methodology and requested that we launch a New York version of the school. They helped fund the project, which began in 1991 in office space in an apartment building in Bay Ridge with a view of the Verrazano Narrows Bridge. The school then moved to the second floor of a synagogue. The location was Flatbush, a community in Brooklyn.

The synagogue, Young Israel of Ocean Parkway, was a yellow brick building with stained glass on one side and a black wrought-iron railing at the entryway. We had the entire top floor and worked with thirty students.

Now there were two Arrowsmith schools—one above the border and one below.

CHAPTER SEVENTEEN

LOST IN SPACE

I remember my father telling me how amazed he was by the two sides of his daughter. "As a child," he once told me, "you could utter profound remarks that showed the level of your insight. But then you would say something else—about how to get from one place to another—that struck me as utterly inane. How could the same brain be capable of both?" And how is it, he had always wondered, that I never knew the route to the family cottage? Straight up Highway 28. So simple. A path I had traveled countless times since childhood, yet I was always confused about what came next en route. The child's eternal question, "Are we there yet?" had a corollary for me: "Where are we?"

There were family stories of me as a young child centered around my getting lost in people's homes during large holiday gatherings, often not to be found for hours. Each person assumed I was with someone else, when in reality I was lost in a maze of rooms, frozen in fear, not knowing how to get back to the gathering. I had no explanation then for this state of affairs, but now I do. I had a specific neurological weakness: spatial awareness (a lack thereof). There is a standard spatial relations test to measure an aspect of this sense, and when I tested myself in May 1978, I ranked in the 10th percentile. Ninety percent of the population was better at processing spatial information than I was. These were the days when I was madly reading Luria's work, learning just what my brain deficits were, devising exercises for them, and trying them on myself, including an exercise for spatial awareness. When I retested myself in March 1980, I saw that I had leaped 80 points, into the 90th percentile.

Before doing the brain exercise that sharpened this area of my brain, this is how the deficit played out in my life for almost thirty years.

I used to feel I would be as successful crossing the street with my eyes closed as open or by making a random dash. How far away is that car? How fast is it coming? If I dash across now, will I be hit? Or do I have plenty of time to get across? I simply could not tell.

Until I went to university, I thought islands floated on top of the water. Only when I did an experiment with nursery school children did I learn this was not the case. I could not visualize the land of the island under the surface of the lake.

Going to a new place always made me fretful and anxious. I was terrified of getting lost and always gave myself lots of extra time ("lost time," I called it) when going anywhere. I just accepted that travel meant having to ask directions repeatedly and progressively until I arrived at my destination. I used to get into my car and point it in the direction I wanted to go. I had mental pictures of my starting point and my end point—and a black hole in between. If I had been to the location before, I would navigate by landmarks and drive slowly enough to use them. I am exceptionally strong in the part of my brain that recognizes objects, and this saved me many times from getting hopelessly lost in familiar neighborhoods. This compensation was not helpful when going to a new location for which I had no picture to guide me of the salient features along the route.

In high school at the beginning of each year, I was assigned a locker. I was always afraid I'd never be able to find my way back to my locker, so I'd routinely enter by the door nearest the lockers. If I came in any other way, I wouldn't be able to find my locker until the new route became familiar by hard repetition.

I was unable to read a map and had no sense of north, south, east, and west. I was very happy when the CN Tower was built in Toronto because I could use it to orient myself. If someone gave me a series of oral directions, I could not translate these into a mental map. I preferred written directions in the form of instructions (go to Elm Street, turn right, go to the first stoplight) over a map. With much effort, I could memorize a route, but retracing my steps, which involves reversing a spatial map to get back to the starting point, was exceedingly difficult, and a detour would easily throw me off.

In a mall, posted maps were of little help to me. Relying on my strength in object recognition, I could imagine the look of the store I wanted, but not that store as part of the mall's overall layout.

I was very poor at building models of any kind. I couldn't draw. I had elaborate pictures in my head but couldn't translate those images into a series of lines on paper. In my first year at university I took a design course, thinking it would be a respite from my science courses; I scraped by with a 60 percent average. When I took photographs, I always cut off someone's head or body or would have everything squished to one side of the frame.

When doing my master's degree in school psychology, I had to learn how to administer the Wechsler Intelligence Tests, including a subtest called Block Design, which confounded me. It involved looking at a drawn pattern and then recreating it using a series of plastic blocks—some white, some red, some both colors. I could not rotate the blocks in my head to create the pattern, so I reverted to the trial-and-error approach: physically rotating each block numerous times and hoping it would match the design. I scored well below average.

The same was true with rearranging furniture at school or at home or when helping my mother rearrange plants in her garden. This was back-breaking work for me—again, because I had to physically rearrange items rather than move them around in my head. My brother Donald, on the other hand, is exceptionally good spatially. He would tell me how he wanted to rearrange desks in a classroom and could imagine five different combinations; I could never imagine even one.

Negotiating a path through a crowded restaurant, passing a car, skiing down a hill: all were a challenge for me because I couldn't plan the moves in my head before executing them. The inability to plan and anticipate spatial moves was panic inducing.

Riding a bike as a child, I would misjudge the distance between my bike and another, with predictable results. When sewing clothes, I couldn't translate the pattern on paper—which is, after all, a map—into a three-dimensional object. I'd have to rip out several seams after putting the front on upside down or the two right sides together or after marrying an inside piece to an outside one. I took it as a given that I would spend twice as long making anything as others in my sewing class.

I could not visualize three-dimensional molecules in chemistry. I was not good at constructing geometric figures. I hated geometry.

Backing up the boat at the cottage or the car in the driveway was always eventful for me. Before reversing, you have to imagine yourself going back-ward through space, and it was hard enough for me to imagine myself going forward through space.

When I put an item of clothing into a closet or chest of drawers, it disappeared. I couldn't hold a picture in my head of items in a drawer; that would require creating a map. Once the objects were put away, they were gone from my mind. I would frequently put things away and not remember where I had put them. My favorite bathing suit disappeared forty years ago. A parallel universe seemed to have swallowed up my possessions.

In my room, I had to have my clothes in full view. Closet and entry doors had hooks on which I hung some clothes, but I also had open shelves, with items of clothing hanging on the sides. Finally, I had clothes folded in piles on the bedroom floor. I was "a pile person," and I would go through those piles to find the one thing I wanted. If I couldn't see my clothes, I didn't know that I owned them.

The list of things I could not do was long.

I fondly remember watching a short, squat character named Mr. Magoo in cartoons in the 1960s. An elderly, wealthy, and blissfully unaware gent who was also notoriously nearsighted, he was always getting lost and into scrapes, but impossible good luck invariably got him and his car safely home, whereupon he would utter his trademark, "Oh, Magoo, you've done it again!"

Imagine getting lost every day. Unlike Mr. Magoo, your eyesight is fine. It's your brain's spatial navigation system that is failing you.

An article in 2008 in *Science Daily* magazine was entitled "Getting Lost: A Newly Discovered Developmental Brain Disorder." There's nothing new about getting lost because of a brain disorder. Since 1978, I have worked with people living with this problem. There is now an understanding that people who get lost are not just inattentive or absent-minded; they have a cognitive impairment.

Which part of the brain governs spatial reasoning and navigation? Current research points to a network of areas, cortical and subcortical. The hippocampus, in the latter area, is a paired structure with mirrored halves on either side of the brain. Interestingly, in Alzheimer's patients, the hippocampus is one of the first regions of the brain to be damaged, and getting lost is one of the early signs of the disease. A cortical region, the parietal lobe (specifically the posterior portion) is also involved in processing spatial information and is activated in tasks involving charting a route from one location to another. More research is now being done on the neural mechanisms underlying how we navigate our way through space.

Eleanor Maguire, a professor of cognitive neuroscience at University

College London, has studied the brains of London cab drivers, who spend approximately two years learning 320 routes encompassing 25,000 streets before being licensed to drive a cab in that city. Compared to a control group without this experience, the drivers' brains were structurally different—specifically, in the posterior hippocampi, an area that plays a critical role in spatial memory. The brains of the cabbies had changed as a result of their environmental demands.

Memory champions, "mental athletes" who strive to memorize in under two minutes the order of a shuffled deck of cards, use a technique called *method of loci*. This involves taking a "mental walk" and visualizing details to be remembered as points along a route and then mentally retracing the route as they recall the cards. Given that this mental walk involves creating spatial maps, the same areas involved in spatial navigation are activated (the hippocampus and the posterior parietal cortex).

Here are two experiments to get a sense of this deficit. Think about your current living space. Take a blank piece of paper and draw the floor plan. After completing the drawing, walk through your home and compare the plan with the reality. Are the rooms drawn in correct spatial relationship to one another? Are the dimensions in proportion? Is anything missing?

Ask someone to read aloud the following directions to you, and as you listen, try creating a map in your head: "Imagine you are standing outside your home. Now make a left turn and walk to the first corner on your street. At the corner go right. Walk two blocks, and make a left. Then go one block and make another left." Did you stop listening after the first turn? Do you have a map drawn in your head? Is it complete with all the turns, or is it fragmented or nonexistent? Can you use the map you have drawn in your mind to retrace your steps to get back home?

Creating spatial maps is a complex process, one that allows us to create internal representations or mental maps of external space. This process allows you both to imagine a series of moves through space and to create a map in your head before executing those moves. If you cannot create spatial maps, it is difficult for you to imagine how to organize objects in space (such as moving furniture) without physically rearranging those objects. Spatial awareness lets you create maps of where you have placed objects in space, allowing you to imagine, for example, what is in your closets.

For someone with a spatial deficit, out of sight is out of mind. For this person, once something is put in a drawer or cupboard, it no longer exists.

These individuals cannot imagine their way into three-dimensional space. Their map-making facility is out of order, so perhaps the best way to explain how this deficit works is to consider two individuals with exceptional map-making skills: hockey players Wayne Gretzky and Mario Lemieux.

Mario Lemieux once likened hockey to a game of chess, but he was the master, several moves ahead of the opposition. "Before I get the puck," he said, "I look where the players are and try to determine where they will be after. . . . It's easy." What Lemieux is describing is his ability to map the field of players and then, even more impressive, almost instantaneously remap the field with the potential movements of every player. Most professional athletes playing the game, even at this elite level and at their speed, might catch one or maybe two players who are open, just fleetingly, and pass them the puck. But Gretzky and Lemieux viewed the ice as if from on high: in their map of the ice, they would know where all ten players were at any given time.

As you can imagine, chess is a major challenge for someone with a spatial deficit. That's because playing this game is in part an elaborate map-making exercise, requiring that the player ponder possible paths for each piece: the diagonal route of the bishop, the linear motion of the rook, the L-shaped trajectory of the knight, the freedom of the queen to go in any direction. And every time a player makes a move, the map changes.

Someone with a spatial deficit, moreover, cannot mentally eyeball the distance between herself and a person across the room, estimate the size of an object in that room, or draw a map of what should be a familiar space— like the floor plan of her own house.

Driving a car, sewing a dress from a pattern, playing checkers, doing a jigsaw puzzle, hitting a baseball: all pose major challenges to those who have this deficit. We once had a student who froze on the stairs when he saw his mother coming toward him while holding a vacuum cleaner. He literally could not imagine how they could pass one another without calamity. Burdened with this brain deficit, some people become phobic and housebound, for distressing experience has taught them that the streets offer one likely outcome: getting hopelessly lost. Crossing the street for these individuals invites panic, and the usual strategy is to search for a stoplight or find another pedestrian to follow.

A spatially challenged person is often a pile person, as I was for some thirty years. When I was a girl in the 1950s, a rainy Sunday afternoon sometimes meant watching movies on television. And I remember one from the 1930s, starring W. C. Fields, in which he played a secretary—but an eccentric (and spatially challenged) one. His desk was piled so high with paper that you could hardly see him behind it. In one memorable scene, his boss asks for "the Johnson file," to which Fields replies in that high nasal voice of his, "Ah yes, the Johnson file." And he plunges his hand deep into the morass of paper and plucks out—the Johnson file.

Pile people organize their space as Fields did. Unable to create visual maps, they must keep everything in sight. To put an object in a drawer or filing cabinet is to lose it.

We once assessed an accountant who had a spatial reasoning problem. Much like W. C. Fields, he kept his papers atop his desk, but he could always find what he was looking for. When this accountant was out of town once and his firm urgently needed a certain document, it took four other accountants hours to locate it in the mountain of paper on his desk. Later in the year, there was a break-in at the accountant's company, and when the police entered this man's office they were certain vandals had been at work. They didn't know that his desk always looked that way.

The worst jobs you could choose if you have this deficit are interior design, dentistry, and mover. An interior designer must be able to hold in his or her mind a visual of a room's dimensions and, with it, the assorted sizes and shapes of furniture in a variety of combinations. Likewise, a dentist mapping the location of a tooth to be filled from an X-ray is also using this cognitive area. Movers, in the same manner, must be able to rotate furniture in their heads to determine how best to angle the pieces when navigating narrow hallways and staircases. Some people simply have a talent for packing the trunk of a car in the most efficient manner. They have a gift for the spatial.

"Some assembly required." Those words, typically printed on the outside of cardboard boxes housing the various parts of barbecue grills, toys, and household goods, strike fear into the hearts of anyone with a spatial deficit. Andrea Peirson has the spatial deficit, and those words call up old memories:

When I worked at a record store in my youth, I would occasionally have to build a cardboard display case to house CDs and cassettes. This was always a source of frustration since the instructions that accompanied the flattened display were a diagram or a map and just had arrows and letters to indicate ("insert tab A into slot B"). I could not understand the diagram, and whenever I tried to build the display, it would never work out right. Often the tabs would end up in the wrong slots and tabs that were supposed to be hidden inside the display case would end up showing on the outside. The display case didn't look nice, but it also meant that the CDs and cassettes wouldn't fit in properly. Finally, the display would almost always end up ripped and then subsequently taped, and/or looking lopsided from my wrestling with the whole contraption. Needless to say, when I became manager of the store, I delegated this task to the boys, who seemed to find it easy and enjoyable.

Males, in fact, typically perform better than women on tasks requiring that three-dimensional objects be mentally rotated.

Video games that involve navigating through mazes are difficult for people with a spatial problem. Several years ago, I was curious to know why a group of my high school students, who didn't have a spatial weakness, always wanted to do the spatial exercise on Friday just before lunch. I soon learned that they went to a nearby video arcade at noon, and they had discovered that doing the spatial exercise prior to playing improved their scores significantly.

My eldest brother, Alex, has (as I had) a spatial deficit (he also has the kinesthetic deficit that I had for almost three decades). You may wonder why he doesn't do the cognitive exercises to address both issues. He is typical of adults who may have one or two neurological weaknesses and have learned to live with them.

Here is Alex's world as seen through the filter of his combined spatial/kinesthetic deficit:

I can't tell my right from my left without thinking about it, and even then it's difficult. That's the spatial deficit. And by the time you've got it figured out, you've run into something. That's the kinesthetic deficit.

It took me a while to realize that not everyone else had this spatial difficulty. I was special. When I was still wilderness canoeing, especially in whitewater situations, where the paddler in the bow and the paddler in the stern must constantly communicate to avoid tipping, not knowing left from right posed a challenge. I couldn't yell out "Left!" or "Right!" so I would yell out, "Near shore!" or "Far shore!" and unless one was in the exact center of the river, this worked fairly well. As a further difficulty, I find it hard to visualize the shapes of objects in 3D, especially if they have to be rotated mentally. This was a big problem for me in university, as I was particularly interested in organic chemistry, where the 3D structure of molecules is very important. I got around this difficulty by repeatedly working with physical models until I could visualize fairly well how they appeared as they were rotated. I also got good at drawing structures in a 3D representation.

Fortunately, I have a very good visual memory for objects, so I can use landmarks to find my way, though I remember one conference that I attended where the hotel was heavily mirrored—especially in the lobby area and where the elevators were situated. I had a very difficult time sorting out where to go once I got off the elevators and would more often than not wander the wrong way until I regained my bearings.

Coming out of underground subway stations and heading off in the right direction is difficult for me. I also have difficulty interpreting direction signs on the highway, especially where several exits and entrances are involved.

Claire Shapiro left her New York City home at the age of eighteen to spend two years at the Toronto Arrowsmith School addressing half a dozen neurological deficits, including spatial reasoning. All her life, Claire had been misunderstood. She was smart, so why was she always getting turned around in the New York subway system even in her late teens when ten-year-olds navigated it easily?

Before that ten-year-old gets on the subway and travels, say, from the Bronx to Manhattan, she automatically and nearly subconsciously deploys part of her brain to create a map. If she goes the same way repeatedly, this process is effortless. For Claire, that process didn't exist. Even creating the simplest "maps," such as a table setting or loading the dishwasher, was beyond her.

"I had lived in New York most of my life," she told me. "It was an abomination that I couldn't get around on the subway. It's not that I had a bad memory. I didn't. I just couldn't hold the map of the route in my head." Claire has a good memory for auditory information, but a memory for spatial maps was beyond her.

Once when she went by herself on the subway (prior to doing the cognitive work to address her spatial problem), she became disoriented when she missed her stop. Her panic must have been apparent to others in the car because many of them started giving her directions.

"Honey, you got to go here, you got to go there," they were saying to her. "So this," she says, "is what it sounded like to me, 'Blah, blah, blah, blah, 75th, 72nd, 85th, up, down, blah blahblah.'" Claire tried calling her mother for help, but her cell phone wouldn't work below ground, and she was almost paralyzed with anxiety.

Until she started doing the exercises to combat her spatial problem, Claire was also intimidated by Toronto's subways. The cognitive exercise Claire did to address this problem involved following pathways within a spatial configuration until her performance was accurate and automatic. This is a tracing exercise, like several others in the Arrowsmith armamentarium, and like so many of those others, it looks simple. For someone such as Claire, this was very difficult. This is, in effect, a mapping exercise that taxes the right hemisphere of the brain.

Eventually the ability to create mental maps came to Claire and, with it, trust in that ability. Even if she got lost, she now had the confidence to ask a stranger, knowing that she could follow instructions and create a fresh map in her head.

The little girl who was always getting lost in her two-floor elementary school and who got from class to class only by following the pack is now a young woman in college. When asked to go to a certain spot in, say, a faculty building, Claire can actually picture where that is and how to get there.

What was once a weakness has now become a strength. In her circle of housemates at college, Claire reports, "I have been able to fix things without calling people to help me, or giving up right away and ending up anxious and angry. I have become known for organizing cupboards and the fridge. I can see where things will fit usually in one or two tries. I tried to organize the fridge very early in the morning when I was still tired, and I found—to my great joy—that it was still easy."

Over the years, I have seen how fatigue affects students with learning disabilities. When we are tired, we may be able to continue to perform tasks that are automatic and routine. But any task requiring effort, such as one needing a contribution from a deficit area of the brain, will become even more difficult when fatigue is a factor. Claire was delighted that she could rearrange the refrigerator without being fully alert, as this demonstrated to her that she could do it effortlessly.

Claire says she retains vestiges of her particular learning-disabled view of the world. "It's an interesting way of looking at the world," she says. "But it's not something you want to be doing for the rest of your life. You don't want to be lost your whole life." Today, she repeats in her head the phrase "You know where you're going" as if she still can't get over that fact and how good it feels.

Today, like Claire, I navigate my world with confidence, and I waste much less time trying to find things or find my way. All that time I spent getting lost is now available for other activities.

I'm much more comfortable driving now and actually enjoy it. As I'm driving down the Gardiner Expressway in Toronto, I can anticipate several moves ahead. I can read maps effortlessly and create maps in my head from verbal directions. I don't get lost any more or fret about getting lost, and I enjoy traveling. I can even plan the most efficient route. I prefer using maps to a global positioning system, as I actually like holding the visual map in my head.

I can visualize in three-dimensional space. When "some assembly" is required to make, say, a cabinet, I can go from the pictorial representation of the diagram and build the cabinet correctly the first time. If something is put away in a drawer, I can visualize inside the drawer to see what's there. I can organize my space well, put items away in an organized fashion in filing cabinets or closets, and locate them easily later on. I am no longer a pile person.

I'm better able to cook a meal. My kitchen working space is more organized, and I use it more efficiently. In the past, I would have all the ingredients and utensils out on the counter, and my mother would be horrified at the mess I made. My kitchen looked like my desk.

What I now know is that improvements in spatial functioning pay immediate dividends: less time is wasted, and life is more organized and runs more efficiently. I redecorated my living and dining rooms and could imagine in my head the best arrangement for all the furniture before positioning it in the room. I had to move my study into a room half the size and was able to use the space so efficiently I had space left over.

I have become a passionate gardener and can hold a map of my garden and move the plants around, all in my head, to create the most pleasing spatial arrangement. I also do collage art and can imagine how the various elements might best combine. My photography has improved immensely—the compositions are balanced. I do not lose things; friends call me as if I were the lost and found department. If I was there to witness a friend place her car keys in her purse hung on the coat tree by the front door, I can call up that spatial map.

For half my life, there were no maps, only the panic, anxiety, and embarrassment that went with constantly getting lost. What a joy it is to have maps in my head and to be able to rely on them.

CHAPTER EIGHTEEN

DRAWING A BLANK

Memory is intricately tied to identity; we are a product of our own experiences. What we perceive is shaped by what we have perceived before; what we learn is bootstrapped on past learning. Amnesia seems to many so horrifying because it robs us of our own autobiography, and thus, it seems, ourselves. If on no other ground, most Americans are joined in our shared desire to improve the curious, elusive faculty we call "memory."

—ALEXANDRA HOROWITZ,
professor of psychology, Barnard College

Annette Goodman has a degree in sociology—or is it industrial psychology? Not long ago, a friend had to gently remind her that her bachelor's degree is in sociology, but her master's degree is in industrial psychology. She has a memory-for-information deficit.

I think of Annette and I think of an hourglass. Invert it, and that upper chamber is momentarily filled with sand, but it isn't long before the sand is all gone and that chamber is empty once more. Her deficit plagues her. She can't talk politics, for about all she can safely say is that Democrats are liberal and Republicans are conservative. Hobbies are out of the question because the amount of effort involved in learning a new skill (such as gardening, which requires remembering facts about plants and planting) would be overwhelming. Unable to store facts in her head, she can enjoy a book while reading it, but then it fades. Stand-up comedy has no appeal because she can't relate to all the cultural references. Likewise, large gatherings hold a particular terror for her because she can't remember names. Annette often worries that a conversation will lead to a question she can't answer and a shutting down. She talks about "the loss of joy associated with speaking freely and spontaneously." What's happening at work and in her children's lives are the two topics she can safely and happily discuss. And when she does remember what she's supposed to, she does not trust her memory, given its notoriously poor record.

Annette feels quite childlike in a restaurant when she asks her husband to order for her. There may be a particular dish she loves, but neither a description of the ingredients nor the name will ring a bell for her. With the memory-for-information deficit, there are no bells to ring.

Studying in advance for an exam is fruitless; cramming the entire night before is the only option. Two tests on the same day pose a huge challenge. Finals mean not just reviewing material previously learned but relearning and memorizing all over again the entire year's material. Twenty-four hours later, it's all gone—again. Having friends over for dinner, you might want to cook or buy their favorite foods and drinks. But if you have this deficit, you can't remember what they like to eat and drink, and you call them apologetically, asking them to refresh your memory and hoping they're not insulted.

"Some things I don't care so much about," Annette told me. "It would be nice to talk about politics or literature, but then there are things that matter more—like when my children say to me, 'What were my first words?' Or, 'When did I walk?' Or, 'What was I like as a kid?' And I'm sitting there, stuck. No matter what you say, they're going to figure, 'Well, Mommy didn't care,' right? I didn't care enough to remember, or didn't notice. Their whole childhoods are a blur."

Eric Kandel, the author of *In Search of Memory,* underlines the critical role of remembering. "Memory," he says "is the glue that binds our mental life together. It allows you to have continuity in your life."

There is no one type of memory. There is a memory for faces, one for objects, one for written motor plans, one for steps in a process, one for phonemic pronunciation, one for spatial maps and patterns, one for body movements, and there is semantic memory for concepts, to name a few. Each type depends on the functioning of different cortical areas within its neural networks. Anthony J. Greene, a professor of psychology at the University of Wisconsin–Milwaukee, where he operates a learning and memory lab, contributed to a special report on memory in the July–August 2010 issue of *Scientific American Mind.* "Memory is dispersed," he wrote, "forming in the regions of the brain responsible for language, vision, hearing, emotion and other functions."

When we work at committing something to memory, the more areas we can recruit to support the learning and retention, the more likely we will remember. Greene explains this process: "It means that learning and memory arise from changes in neurons as they connect to and communicate with other neurons. . . . When a memory forms in the brain, it alters the connections between nerve cells. . . . New synapses form when repetition reinforces an association, creating a memory as neurons that fire together wire together."

In this chapter, we are looking at what I call *memory for information*. This form of remembering details and facts is primarily an auditory process. This type of memory critically supports us in carrying out our myriad daily tasks: recalling conversations, remembering appointments and errands to run, retaining instructions in the classroom and workplace, and so on. People with good functioning here take this ability for granted. For them, this kind of memory operates in the background, with little conscious thought.

What happens, then, when this type of memory is not working?

Imagine a series of sieves with different-size holes—some the size of an olive, some the size of a pin. Now think of all that you hear and want to remember over the course of a day—instructions from your employer, facts of interest on the radio, details of a procedure, items to purchase—and imagine them flowing into the sieve. Someone with a severe problem in this form of memory has large holes through which the information leaks. They retain only bits and pieces. Someone with a mild problem has smaller holes, so more—though not all—is retained. Those with good functioning have no holes and can retain what they want, discard what they don't need, and build up a stored bank of factual information.

Think back to two weeks ago from today. Can you remember what you had for dinner? Now think about what you had for dinner last night. Compare the effort it took to retrieve both details; the former task was clearly the harder one. Those with a problem in this area face that kind of struggle when they try to call up factual information from memory. Either they never retained the material, or it takes a huge effort to recover even fragments of the memory.

I remember a man who had notes written all over his hands, so poor was his memory. Individuals with memory challenges sometimes make lists, sometimes only two items long, of things they must do in the next hour. Others have sticky notes in every corner of their house and office. And then there was the pilot who could remember only two oral instructions at a time and so had to have the traffic controller repeat the instructions multiple times before he could retain them.

In *Higher Cortical Functions in Man,* Luria describes a memory problem resulting from damage to the left temporal lobe in which a person cannot repeat a series of short phrases after hearing them. Individuals with this problem, he notes, have difficulty imprinting, storing, and recalling information they hear.

In 1978, a nineteen-year-old high school graduate came to me describing his memory as a leaky sieve, an image I have since used to describe this deficit. He had not applied to college because he could not imagine being able to remember all the content required. I set out to investigate what was going wrong with this young man's memory function. I asked him to listen repeatedly to popular songs with lyrics, and here is what I found fascinating: after listening to the same song five times, he could hum the melody perfectly but could not remember the lyrics. When I repeated this with songs that had more complex melodies and simpler lyrics, the same phenomenon occurred: he got only snippets of the lyrics.

When I, as Luria described, asked him to repeat short phrases, he was stymied. He could not repeat two short phrases back to me. I then selected songs and graded them on a continuum of easier (shorter, with more repeated phrasing) to complex (longer, with varied phrasing). His assignment was to listen over and over and over again to the simpler songs until he could repeat the lyrics, and then repeat the same process progressing to the songs with the more complex lyrics. After several months of this practice, his memory for short phrases doubled. No more leaky sieve. Thus was born the Arrowsmith cognitive exercise for the memory-for-information deficit.

Clearly there is a genetic component with cognitive deficits. Avital, Annette Goodman's daughter, inherited her mother's memory-for-information deficit; indeed, hers was worse than her mother's.

Some parents may fail to recognize some daunting possibilities: that their child's learning disability may be much more severe than their own, or that their child has inherited different deficits from the mother *and* the father and that the end result may be even worse than either can imagine. Such parents may have struggled in school with learning challenges but are now functional and successful, leading them to believe, mistakenly I think, that their child will "make it," just as they did.

Avital would not have made it. Given the severity of her memory deficit, she would not have been functional as an adult had this deficit not been addressed.

Annette was concerned about her three-year-old daughter's inability to remember even basic things, such as colors. The nursery school principal interviewed Avital and presented a color palette to the child, who had already developed defenses around this deficit.

Asked to identify a certain color, she said, "That's not my favorite color."

Annette looks back on that incident and offers a wry remark: "Three years old, and she's already aware that she has learning problems, and she has devised a system of trading on her charm to hide it."

The list of things Avital could not remember was long and frightening. Even by the age of seven, she still could not remember the city she lived in, her address, her telephone number, or her last name.

"We didn't even tell her she had a second name for years," says Annette, "because we wanted to focus on the last name. Recently she asked us, 'Why didn't you ever tell me I had another name?'"

Avital's parents were terrified of her getting lost, for she would have been completely vulnerable. She could play with a friend for hours and afterward not remember the other child's name. She once spent a week with me and could not keep in her head the names of the two cats she played with and clearly adored.

Her teacher's name. The days of the week. The months of the year. All this basic information would not register on Avital's brain.

She wasn't functioning in school, and yet she was clearly a smart little girl. Her defenses were many and strategic. She used the fact that she was adorable to win friends and try to outmaneuver her parents. She would distract her tutors with disarming stories to avoid the work, frustrating them constantly. And at the end of every day, Avital was visibly and understandably exhausted.

At home, the memory deficit shut her out of family games, such as the Goodman tradition of playing board games or poker with chips. How could she play the latter when she couldn't remember the value of a red poker chip versus blue, or the former when she couldn't remember the rules?

In first grade, Avital was tested and found to have a wide range of neurological weaknesses, most in the severe range. She was placed in an Arrowsmith classroom, and day after day and week after week, she did the cognitive exercise meant to build her auditory memory. To stimulate that area of the brain, she was asked to memorize poems following a specific procedure. As is the goal of all the cognitive programs, the end result was not just that Avital got better at reciting poetry; she could now retain information she heard in class and at home. The stimulation had strengthened the neural underpinnings of this form of memory. And as with all other Arrowsmith exercises, the work began simply and got more complex. The quicker her recall, the faster Avital moved to the next level. Annette began to see change in her daughter just three months into the exercises, and that improvement continued, slowly but surely. Annette tells a story to illustrate her daughter's memory deficit before and after cognitive treatment.

"Years ago," she told me, "Avital stood at the back of the stage during school plays and just moved her mouth. She had to remember two words— and she couldn't." Avital's brain was not able to make the connections, to forge the neural pathways necessary to remember common words, never mind instructions or dialogue. "Two years into the Arrowsmith Program," Annette says, "I went to the first play where she had two lines, and I cried seeing her remember her lines. This year, in sixth grade, she auditioned so excitedly for the school play and talked about which parts she preferred and is working on her lines with full confidence. During rehearsals, she not only remembered her lines, but she helped her friends when they got stuck. Avital knew almost everyone else's lines by heart."

Avital can plan her own birthday party now and keep in her head what needs to be purchased. She plays poker with her older siblings and cousins, capably keeping pace with any adult in the game, as well as Clue (remembering the names of characters, rooms, and weapons in that murder mystery board game). As Annette puts it, "She's part of the family now."

Finally, Avital is more relaxed. There's no need now for those elaborate defenses and strategies that consumed so much of her energy. She's now considered the "go-to person" in the family—the one who will gently remind her mother whom they borrowed ski equipment from last year or the name of the hotel they had booked two weeks earlier.

"We were playing 'geography' with her recently," Annette told me, "and instead of using locales we used food. So as one person named a food, the next person was responsible for naming a food that started with the last letter of the prior food. Avital was faster at generating foods than I was and whispered words such as *tilapia* and *horseradish* in my ear to help me out when I got stuck. And best of all," says Annette, "I no longer worry that I will have to repeat elementary and high school with my daughter, reteaching her all the material. She's independent now."

Amber, a young mother, is a talented songwriter and musician who performs on guitar. She sings and writes her own songs, and to date she has written and is in the process of recording seventeen songs. But when she came to us as a twelve-year-old girl in the mid-1980s, her mother, Mary Arlington, reported that when she would ask Amber to go into the basement and fetch some things, the maximum she would ask for was two. Among her many learning disabilities, Amber had a severe memory-for-information deficit.

"Though her intentions were good," her mother recalls, "she couldn't remember being told to do things. I'd say, 'Amber, clean your room; then go do the dishes.' She'd forget because in the time it took to perform the first task, she had forgotten the second. It would have been easy to call her lazy and unresponsive," says Mary.

Amber has a high IQ (in the very superior range), so there was a huge disconnect between what Amber could actually do and what she was in theory capable of. She was, of course, immensely frustrated, and when other students bothered her in class, she was inclined either to beat them up or run away from school. Amber's mother, a medical doctor with training in psychiatry, had always suspected a learning dysfunction, and Arrowsmith testing revealed precisely that.

After two years at Arrowsmith, Mary and Amber both reported that old mother-daughter antagonisms had subsided, and each thought the other was now more fun to be around.

Parents of a child with this deficit are often convinced the son or daughter is irresponsible, lazy, or stubborn because the child doesn't do what he or she is asked to do and forgets to perform household chores or fails to follow a series of instructions. Parents may mistakenly believe that the child remembers only what he wants to remember. The problem, as in Amber's case, is cognitive limitation, not emotional resistance. The misunderstanding sets up a negative dynamic between parent and child, but that dynamic ameliorates once the deficit is addressed.

Amber sang in those days, but was never the lead singer. The potential for forgetting lyrics on stage, the risk of embarrassment, was too great. Now she sings lead with confidence. Her once-faulty memory became reliable and twenty-five years later remains reliable.

"How amazing," Amber told me, "that as a kid I could never remember lyrics. Now, among my musician friends I have a nickname—'the vault.' 'Call Amber,' they say, 'she knows the lyrics.' Not only do I know the ones I wrote, I know what other people write. When I'm in a performance, a high-stress situation, I never forget the lyrics. The guitar may fail, but I don't."

Amber's newfound ability to write lyrics calls to mind a point I'm often asked about. Will strengthening any one brain area "rob" a student of a strength in another area? There is a myth that students will lose their gifts if they address a learning dysfunction; that somehow a "gain" in a weak part of the brain necessarily means a "loss" in an area of strength. In fact, as cognitive areas are made stronger, new connections are made within and between brain areas. Therefore, an area of talent can now flourish as it is no longer hampered by the interference of a learning dysfunction.

Maureen O'Toole vividly describes her daughter Kathleen's anguish over her learning disabilities. "She cried and cried and cried," Maureen told me. "Every single night she cried. The whole house shook with her tears."

Arrowsmith testing revealed that Kathleen, then in seventh grade in a small-town school, had a long list of deficits, including the memory-for-information deficit. Nevertheless, she was smart and extremely capable.

"She's a put-together girl," is how Maureen describes her. "A great kid. She's responsible, supermature. I trusted her with my little kids. She was great with them."

But there was a flip side to this picture of an unusually mature girl, and it was eating at Kathleen. "She felt totally stupid," said Maureen, herself a teacher. "Kids are in school six hours a day. They come home, they have homework, so it's really their life. And if they're not successful in school, they really feel totally negative about themselves. Their self-image is just totally knocked."

Kathleen's memory deficit meant that one of her great fears in class was that a teacher would erase a blackboard before she had caught up. She was at a double disadvantage: two neurological weaknesses converged to make note taking almost impossible for her. She had a handwriting challenge from a motor symbol sequencing deficit, meaning she could not write quickly enough to copy the notes from the board before the teacher erased them. And she had a memory-for-information deficit, meaning she could not remember what the teacher had said. The notes on the board were her only point of reference, and once they were erased, that lesson disappeared with it.

The lowest point of Kathleen's struggle occurred in sixth grade. Maureen would study with her night after night, but the next day Kathleen would have forgotten it all. She just could not remember things. Kathleen would be sent off to run two errands: take a jacket to the cleaners and buy a carton of milk. And she'd come back with the milk, the jacket still in hand. Maureen would ask Kathleen to watch something in the oven and take it out in half an hour, then come home to burned cornbread.

After two years in an Arrowsmith Program, that's all changed. No more burned cake in the O'Toole household. No more vacant stares when Kathleen is asked what she learned that day in school. And, like many other adolescents, Kathleen sometimes forgets what her mother tells her to do, but now it comes back to her later.

On a statewide biology exam, one that severely taxes memory, Kathleen got a mark of 95.

"Things are good," she says. "Very good." She will not soon forget how good it feels to remember.

CHAPTER NINETEEN

SEEING AND NOT SEEING

1994

The move to take the Arrowsmith concept to New York started to come undone in the early part of the 1990s. I was flying back and forth between Toronto and New York, so much so that neither city felt like home.

The decision was made to wind down the Toronto school in order to focus on the New York program; then it too eventually folded—in no small part because my marriage to Joshua was coming to a tumultuous end. I made the decision to return to Toronto late in 1994. Joshua remained; he died in New York in 2000 of cardiac arrest.

When, as part of research for this book, I was in New York in fall 2010 visiting schools that implement the Arrowsmith Program, talking to teachers and principals and parents and students, I traveled endlessly—from Manhattan to Long Island to New Jersey and the boroughs and towns beyond. At one point, I crossed a bridge that took me close to the cemetery where my husband is buried.

My husband. He was a deeply wounded soul whose gift to me was not denying me my learning disabilities and bringing me to the work of the brilliant Russian scientist Aleksandr Luria, who utterly changed my life. So thank you, Joshua. As I passed the cemetery that day, I offered a prayer. May he rest in peace.

———————

My marriage had endured fourteen years, and when it ended, so did a seventeen-year relationship that had begun in the spring of 1977. I was not the same person at the end of that period as I was at the beginning.

During the first years with Joshua, my reasoning deficit left me particularly vulnerable to manipulation because I could never be certain what people meant. With some people, there was surface meaning and then there was the true agenda. I could not tell the difference.

In the wake of the cognitive exercises, I had to learn for the first time how to use a brain capacity I had never possessed. Like someone gaining sight after many years of blindness, I found the transition exceedingly difficult. The hallmark of the symbol relations deficit is an abiding sense of uncertainty. After twenty-six years of uncertainty and of missing the nuance in people's communication, it was difficult to finally trust my understanding. And the emotional turmoil, the low self-esteem that had developed over my first three decades, lived on during my time with Joshua.

I used to think naively that once one's cognitive deficits were addressed and the learning capacity enhanced, the emotional healing would happen spontaneously. My own experience and that of people I have worked with suggest otherwise.

The dynamics that govern abusive relationships are complex, and I do not mean to diminish all the factors that contribute. My intention here is to underline the role that learning disabilities can play in this troubled dynamic.

In 1995, trying to make sense of my time with Joshua, I wrote:

The increase in the amplitude of his abuse, all done so gradually, so subtly I was unaware of being caged, until he controlled all outside contacts. My perceptions subtly manipulated by his will. My vision circumscribed by his mind. My world defined by him. A form of agnosia, perceiving sensations but being unable to interpret them without his overlay. The agonist binding to the receptor, fitting together, bound, the pathology incomplete without the two interlocking. Like a chemical reaction, setting off a chain of events no longer under conscious control.

If I had set out to write a story about the intersection between the nature of specific learning disabilities, the vulnerability this creates and the impact all this has on one's emotional well-being, I don't think I could have done a better job than describing my own life. Joshua knew my cognitive limitations intimately, and his barbs would deepen my already well-developed sense of inadequacy:

"Why are you so stupid?" he would say.

"Why don't you understand? It is so obvious."

"With your lousy sense of direction, it's a wonder you ever get anywhere."

"You are so clumsy; I can't trust you with anything."

A litany of criticism, each based in a measure of truth. My habitual response was to try harder to get it right. Early on, I had learned to suppress my pain, deny my emotions, and redouble my effort. As Zazetsky so aptly said, "I'll fight on." This was my mantra. Years later, a researcher who interviewed me likened my world to amygdala hell. The amygdala is the brain's threat detector, readying one for flight or fight, and mine was on constant high alert.

Where my learning disability had, in its own particular way, closed me off from the world, now my relationship with my husband did. Who I was permitted to speak with or see became more and more restricted until only he was left.

I knew when I met Joshua that he was emotionally wounded, and in my naiveté and arrogance, I wanted to heal him, to make him right. All my life, that had been my instinct. But by the end of our time together, I was the one in need of healing.

It was clear to me when I fled Joshua that the blinders I wore because of my neurological deficits had gotten me into that relationship. Cognitive exercises had removed those blinders; learning to understand the world and starting to heal had gotten me out. I felt an odd mixture of sadness and relief.

During those years with Joshua, my work with learning-disabled children became my refuge, the place where I put my heart and my soul and all my creative energy. I was, and I remain, grateful for this work, which is healing work.

On November 18, 1994, I left New York for Toronto. I was alone and very much relieved, grateful, actually, to be alive. But my health was very poor, and my spirits were at low ebb.

The notion of work, of making a living, of starting my life over again would have to wait while I recovered, for I had been rattled to the core in every way. Psychologically I was numb. Physically I was in a very poor state after four miscarriages and multiple surgeries for endometriosis. The school in New York was gone, and so was the Toronto school. What equipment remained from the Toronto school had been put into storage. I had come back to nothing—less than nothing. This was a bleak time in my life.

But then the most remarkable thing happened. Helen Fadakis, then and still the bookkeeper at the Toronto school (where her own daughter had been a student), called four key parents of other students and somehow managed to relaunch the school. They paid the storage fees, retrieved the office equipment, and convinced me to get back to work. They found a new space on Yonge Street and got the school going again. These angels as much as said, "There has to be an Arrowsmith school in Toronto. We need it, and others do too."

CHAPTER TWENTY

WHEN 2 + 2 DOES NOT EQUAL 4

"Can you do Addition?" the White Queen asked. "What's one and one and one and one and one and one and one and one and one and one?"

"I don't know," said Alice. "I lost count."

"She can't do Addition," the Red Queen interrupted.

—LEWIS CARROLL, *Through the Looking-Glass (1871)*

Mary Arlington, the Toronto doctor introduced earlier in the book, had difficulty with time and numbers. Imagine being a clinician trying to run a busy medical practice when you don't have a sense of how long a minute is, and the numbers 8, 80, and 800 all mean pretty much the same thing to you.

Mary's quantification deficit was affecting her in many critical ways. For one thing, her finances were in considerable disarray (Did the new furnace cost $800 or $8,000?). For another, she was always running late as she lost track of time and immersed herself during appointments in the tangled lives of patients with severe anxiety disorders, schizophrenia, and bipolar illness. And since she booked her own appointments, patient scheduling was often confused and full of conflicts.

Mary compensated for her inability to make sense of numbers and logic by enlisting her remarkable nonverbal strengths, which led her to develop in the early 1980s what were then considered radical ways (using diet and vitamin supplements) to treat mental illness. She gathered research on this subject over the course of many years, but then didn't know how to collate all the data.

Mary would listen to a message on her answering machine and have to replay it several times before she was able to get down all the digits in the telephone number. This wasn't a memory-for-information issue (as discussed in Chapter 18); this was about memory for numbers.

Two years of doing cognitive exercises at Arrowsmith in the mid-1980s induced a remarkable change. One year into the program, Mary astonished her bank manager by doing loan agreement arithmetic in her head and presenting him with an updated and balanced checkbook. She no longer needed an overdraft account.

Mary is quite effective at math now. As part of treating some patients, she does metabolic assessments involving lots of data (levels of zinc, histamine, cholesterol), and she has developed a flow sheet that tells her at a glance where the problems are.

Mary is no longer chronically late. Even baking a cake is easier for her now: if she has to guess when twenty minutes are up, she's usually within a minute or two. She also came to understand the relativity of numbers. Mary knows that the Los Angeles urban area, with 15 million people, is bigger than that of Toronto, with 6 million. Until her quantification weakness was addressed, she had no sense of the relative magnitude of numbers.

Mary enrolled in the Arrowsmith Program after she'd seen its success with her daughter, Amber. Both mother and daughter had the same problem with numbers.

Mary's first inkling that Amber had difficulty with numbers occurred when her daughter was quite young and having trouble with the multiplication tables. She was also refusing to do math homework. Mary or a teacher would work with Amber on the tables, and Amber would appear to have that knowledge in her head. Then it would simply fade. Mary would spend her summer teaching her daughter the multiplication tables, knowing that the following summer, those lessons would recommence.

These days, Amber, who can now do mileage calculations in her head thanks to firmly entrenched multiplication tables, does her mother's accounts and has recovered tens of thousands of dollars in lost billing.

For both mother and daughter, the number phobia has been put to rest.

I initially labeled a problem with number sense, quantification, and mental calculation as a *supplementary motor deficit*, based on two things: Luria's description (in his book *Human Brain and Psychological Processes*) of a breakdown of mental arithmetic in a young man with a wound in the supplementary motor area of the brain, and an early neuroimaging study in 1978 by the Swedish researcher Niels Lassen and his colleagues, who found that counting in one's head repeatedly from one to twenty activated this same area. More recent studies, however, suggest that this area, while at times showing increased activation during mental calculation, probably plays a supportive role at best.

Our understanding of what brain areas are involved in which mental activities has historically come from two key sources: lesion studies (which functions are lost, or not, when there is damage to a specific area of the brain) and neuroimaging studies (which areas are active when we perform specific tasks). Both sources have their flaws. Data from lesion studies are imprecise since lesions are rarely confined within clear anatomical or functional boundaries. Imaging data tell us what areas become active when specific tasks are performed but not which of these areas are critically involved, which are marginally involved, and which are incidental—that is, triggered by how the task was performed.

Martha Farah, founding director of the University of Pennsylvania's Center for Cognitive Neuroscience, gives an example to illustrate how areas of the brain unrelated to a certain task can nonetheless be activated. If instructed to count the letters in the first word of this sentence and to suppress reading the word, you will find that you automatically read *and* define the word, activating brain areas unrelated to counting. Investigating the number-processing function is not a simple matter. The question researchers ask, and how it is asked, will have an impact on which areas of the brain are activated. Interpreting data from lesion or imaging studies needs to be done with caution.

So if not the supplementary motor area, what other parts of the brain might not be functioning properly in Mary's or Amber's case?

Though our understanding of the neural circuits underlying number processing is incomplete, one brain area, the horizontal intraparietal sulcus (hIPS) in the parietal cortex, is thought to house our number sense: our intuitive understanding of numbers, their magnitude and their relationship to one another.

Stanislas Dehaene, author of *The Number Sense*, calls this area (hIPS) our "quantity circuit." It supports mathematical operations (addition, subtraction, multiplication) that involve manipulating quantity, so a problem here will interfere with all these operations.

"We cannot think about a number," says Dehaene, "without activating this brain area." And this holds true no matter how the number is presented—whether in arabic numerals, words, spoken numbers, or represented as dots or tones. When presented with a numeral, Dehaene says, the brain immediately begins a process to establish that number's relative size. It's as if we have a mental number line that allows us to place and understand numbers in relation to each other.

The left angular gyrus (discussed in Chapter 8) is another brain area involved in number processing. Thought to be involved in mathematical operations requiring language (memorizing addition or multiplication facts), this area also mediates the successful acquisition of broader mathematical knowledge and concepts. Both the hIPS and left angular gyrus are important to our understanding and processing of numbers.

When the hIPS area is not functioning properly, the result is quantity blindness. Brian Butterworth, at the Institute of Cognitive Neuroscience at University College London, says, "We're born with the ability to see the world numerically just as we're born to see the world in colour." But some people are born with a kind of number blindness.

A sense of number is essential for operating competently in the world. Without it, the world of numbers is like a foreign language. To get a sense of this deficit, ask someone to slowly read the following series of numbers to you: 5-1-6-9-3-2-7-4. Can you repeat them in the correct order that you heard them? If so, can you repeat them in reverse?

Now try this exercise with a shorter series: 4-8-1. Note how much easier it was with fewer numbers to remember. Someone with a problem in this area experiences all operations with numbers, no matter how simple, as a struggle.

Characteristically, individuals with this deficit have difficulty performing operations with numbers in their heads ("mental math," educators call it). They struggle even with simple addition in their heads and need external support (finger counting, counting out loud, calculators). But they can't rely on the calculator because they can't estimate if the answer computed is reasonable. And because number has no meaning for people with this deficit, they have difficulty learning addition and multiplication facts.

The word *quantity* comes from the Latin word *quantus*—meaning "how much." A challenge for all individuals with this deficit is knowing how much—of money, distance, or time. To communicate about quantity, we break it down into measurable and equal units based on number: sixty minutes in an hour, one hundred pennies in a dollar. We can memorize these relationships as facts or understand them almost instantly as units of measurement. We know how much of something we have, how much we need, and how much is left to spare. In the case of money, someone with this deficit will wonder: Do I have $10 in my wallet, or $80? If my purchase was worth $9 and I gave the clerk $20, how much should I get back in change?

As with other dysfunctions, people behave differently: some try to compensate, some can't. One person may compensate by always carrying far more money than might ever be required; another may spend very little and save almost everything; another may find herself always out of money, having spent everything and unable to account for where the money went. Unable to determine how much gas is required to go a certain distance, this person runs out of gas on the highway and always relies on someone else to calculate the tip at a restaurant. Counting calories would be similarly problematic, and Sudoku puzzles are out of the question. For someone with this deficit, any job requiring time management—working as a chef, for example—would be a nightmare. Add the specific demands of number sense, and all is literally lost.

Picture what happens when a chef receives multiple orders from a number of wait staff simultaneously. The chef needs to mentally calculate the cooking time of each dish so that every order slated for a particular table is ready at the same time. He or she also needs to mentally calculate the total number of items in all the orders and double or triple ingredients based on the number of identical dishes needed. If the chef is also responsible for ordering ingredients, then he or she must calculate appropriate amounts as the menus change with the seasons and customer tastes. Other jobs that would pose a challenge for anyone not strong in this cognitive area would be stock analyst, actuary, accountant, and statistician.

Individuals with this cognitive weakness have difficulty structuring their time in order to complete a series of tasks—for example, doing what they need to do between waking and leaving home in the morning. They cannot estimate the time required for each task or, as each task progresses, the time remaining. Quantified intervals mean little to someone with this deficit, so if you said you wanted to meet that person a week from Tuesday, she would literally have to run through the days of the week to know how many days away that was.

There are, of course, ways that people can compensate. They can, for example, develop a rigid morning routine from which they dare not stray, but as soon as they encounter a new situation requiring them to use time as a way to structure multiple tasks, they will have difficulty. Some will compensate by leaving an excessive amount of time in which to complete a task at work, possibly by coming in early or staying longer. Others may eventually rush through the task, leaving parts out—for example, rushing to catch a plane and not packing half the clothes needed or leaving home without taking a shower or eating breakfast.

Anyone controlling the financial activity of a business must have a strong mental math capacity because that person must continuously be calculating cash flow, profit, loss, assets, future earnings, current overhead, capital investments. Someone with this deficit would find it impossible to run a business.

Rabbi Eliyahu Teitz is associate head of school of the Jewish Educational Center in Elizabeth, New Jersey. "When our school agreed to host the Arrowsmith Program," he says, "we saw great possibilities in helping children who were not reaching their potential through traditional educational methods. We were willing to see what the program could do." He admitted to some skepticism about the results being reported until the stories started filtering in about his own students. One came from the father of a child who had been struggling with the quantification deficit. The father wrote:

> This past weekend, while walking my son to his friend's house, I started quizzing him: 1 + 1, 2 + 2, 4 + 4, 8 + 8 etc.
> When he answered 16 + 16, I was relieved.
> When he answered 64 + 64, I was impressed.
> By the time we got to 1,024 + 1,024 I think even he realized what he had accomplished. Though he won't acknowledge that it is due to Arrowsmith, I do.

I remember Jennifer, who was sixteen when I first met her. She had been able to learn the alphabet with ease in first grade but not numbers. She had no difficulty learning how to read *Monday, Tuesday, Wednesday,* and the other days in the week, but she had no idea of their chronological order. She had no concept of a calendar, or how far away one date was from another, and she was convinced there were 189 weeks in a year.

Creating a schedule was impossible for her, as she had no sense of how long any activity would take. In order to compensate, her mother used a timer and broke down each task in her morning routine into a specific number of minutes. When the timer went off, Jennifer had to switch over to the next task. This was the only way she could leave the house on time. When asked what she would do with $10, she said she would go to the movies and buy a chocolate bar, and maybe a car. The world of numbers and time had no meaning for her.

Jennifer could not compute addition or subtraction questions in her head or solve math questions; she would make tiny dots on paper that she then counted to calculate the answer.

As she worked over a period of two years on this deficit, numbers came alive for her. Jennifer learned her addition and multiplication facts. She no longer needed to rely on the timer to get out of the house in the morning. In fact, she became excited about beating the timer and then eventually put it away after developing an internal sense of time. And Jennifer was able to do all this: learn how to understand and use a calendar to plan events, open a bank account, and calculate how much of her allowance she could devote to saving and how much to spending.

Numbers at last made sense to her.

Cash in envelopes marked "Rent," "Food," "Bills to Pay." That's how Ann Tulloch managed her own finances from month to month. Doing basic arithmetic, planning a budget, balancing a checkbook: all were beyond her. And until she devised this rudimentary system for managing her money by earmarking it, Ann was perennially short of cash and having to borrow, which put her into debt. Even as an adult, she could not do simple calculations $(8 + 3)$. In 1985, at the age of thirty-three, Ann found her way to Arrowsmith by pure serendipity.

Testing showed that she had five areas of deficit, including the one that led her to put cash into envelopes—a quantification deficit. Ann could not perform mental calculations in her head. Numbers were just symbols on a page with no meaning for her. In school, she would squeak by with a passing grade of about 50, but she finally reached the end of her tether in tenth grade. She was told that even vocational school was inappropriate for her owing to her bad attitude. Her response at the time? "I don't have an attitude problem. I just can't learn anything."

Drawing and design were her twin passions, and the fact that one of her brothers went on to become an aerospace engineer might suggest a talent that ran in the family. But her poor grades meant that art school was at first out of reach.

There followed a series of menial jobs. Filing clerk with an insurance company. Working the concession stand in a cinema. The vice principal who tried to talk Ann out of quitting high school was partially right when he told her, "You have the makings of a Rolls-Royce, but you only put forth the effort of a Volkswagen." He was wrong about her lack of effort but right about her potential. Ann did later enter art college as a mature student, but her crippling deficits ruled out a career as a commercial artist.

When she was living at home, she had seen a police drama on television in which a brother explains that his sister wasn't retarded, as some believed, but that she had a learning disability. For the first time, Ann had an explanation for her inability to learn, but her parents dismissed the possibility. Her problem had a name, but the solution was no closer.

Shy, unable to articulate properly, she had little confidence and low self-esteem. Ann quit her various jobs and worked for two years on a horse farm, a virtual sabbatical from the human race. The horses were grateful to be fed on time and never once called her stupid. Then came the break. By chance, she encountered Tanya Day (an Arrowsmith student introduced in Chapter 10), who shared Ann's passion for horses and who often got a ride with Ann to the stable.

"We would drive up to the stable in the car," Ann remembers, "and she would say, 'I like cars, I like black cars. I like cars, I like black cars.' And that would be about it. But you could tell she was smart. She could remember directions. She just couldn't talk very well. And then she started going to Arrowsmith, and all of a sudden she could talk. It was wonderful." Ann thought, *If they can help Tanya, maybe they can help me.*

So Ann embarked on an almost four-year-long course of exercises to deal with her various cognitive weaknesses. Like so many who are puzzled and disheartened by their lifelong inability to learn despite their intelligence, Ann felt very relieved.

"I felt for the first time in my whole life," she told me, "that somebody finally understood me. I felt, 'I have this learning disability, and now I'm doing something about it.'"

Arrowsmith's cognitive exercise for a deficit in this area involves repeated and progressively more difficult mental calculations such as addition and subtraction. This exercise works the quantification circuit. Accuracy is an essential component of these calculations, which must be done within fractions of a second. And as with all other Arrowsmith exercises, once mastery is obtained at a simple level, the next level adds complexity, placing a greater demand on the brain.

"It was tough," Ann says of the exercises to overcome her deficit. "So frustrating."

In an extraordinary display of sympathy and understanding, Ann's employer came to the school to review the test results and to better understand the problems that beset Ann. As Ann started to make cognitive progress, her employer decided to hire her full time. Secretarial work morphed into managerial work and a substantial increase in income.

A combination of significant deficits had made it difficult for Ann to speak as well as understand conversations. She had become shy and withdrawn to the point that she had allowed others to take advantage of her. As her brain functioning improved, she began to make significant changes in her life. She dropped, for example, a ne'er-do-well boyfriend who had convinced her that he was looking after her when the reverse was true. Similarly, in social situations, Ann stopped allowing herself to be used or "pushed around," as she puts it. Such treatment, for the first time in her life, registered on her—though not always right away. But she found herself going back to someone and saying, "Hey, what'd you do that for?"

"I feel like I have more control in my life," Ann said, summing it up in 1986. "The difference is so incredible. Other people wander around with all this stuff as normal, and they just take it for granted, and for me it's so exciting now. And when I get into a new relationship, it's going to be a different one. Very different. For the first time, I'll be going in with my eyes open."

In 1996, when Ann's father was eighty, Ann decided to speak to him openly and for the first time about her learning disabilities, how they had affected her life and self-esteem. As she was growing up and often in tears over yet more homework she could not do, he would frequently ask her with impatience bordering on anger, "What is *wrong* with you?"

Now he listened and apologized—the only time Ann remembers him apologizing for anything. "I am so sorry," he told his daughter. "I didn't understand." Ann revealed to him the work she had done eleven years earlier at Arrowsmith and how she had held down two jobs to pay for the program. A light went off for her father.

"Our relationship changed," she said. "I was no longer angry, and he could understand what I had gone through. He asked what I had paid for this work and wrote me a check to cover what I had spent. In his way, he wanted to acknowledge my struggles and to support me in what I had done to overcome them."

Finally, Ann's finances are better. She can do math in her head, she no longer makes mistakes on company invoices, she balances her checkbook (if she forgets to enter a transaction in her register, she retains the amount and does it the next day), she knows how much money she has in her wallet since she can keep a running account, and she started a registered retirement savings plan, so she's begun to plan ahead. She is now capable of that kind of financial planning. Ann was also able to budget for, and take, her dream vacation.

No more cash in envelopes.

CHAPTER TWENTY-ONE

IN ONE EAR AND OUT THE OTHER

A nn Tulloch had a quantification deficit, but like so many of us with brain deficits, she had several, including one that I call an *auditory speech discrimination problem*.

Speech processing starts in the superior temporal region (located somewhat behind the ear) in both hemispheres of the brain. This area is involved in discriminating the sounds of speech from other sounds such as noise or music. Then, in the left hemisphere, speech sounds begin to be identified and discriminated one from another—what I call auditory speech discrimination.

At the age of three, a toddler's brain is roughly the same size as an adult's but with nearly twice as many synapses. The child develops, keeping the synapses being actively used and pruning away those not being used. For example, an infant can discriminate speech sounds from all languages, not just those that exist in his native tongue. The brain restructures the auditory networks to become tuned to the sounds regularly heard and loses the ability to discriminate sounds that are not part of that infant's auditory world. This rewiring, which occurs naturally and automatically, increases our brain's efficiency while applying one critical principle: use it or lose it.

For reasons that remain unclear, however, someone with an auditory speech discrimination disorder never gains precision for discriminating speech sounds. Invariably, testing by audiologists will show that a person with this deficit hears normally (that was certainly the case with Ann, who was tested several times). The problem is that owing to an impairment in the brains of these individuals, the word *hog*, for example, can sound like *fog*. Distinguishing like-sounding words such as *fear/hear, doom/tomb,* or *vow/thou* poses a major challenge.

Wepman's Auditory Discrimination Test is designed to measure the ability of children to recognize, through listening, fine differences between like-sounding words. The Wepman test comprises forty pairs of words. The words in each pair are one syllable and of equal length. In ten of the pairs, the words are identical. In the remaining thirty pairs, the words differ slightly. During the test, the listener, who is not allowed to see the examiner's lips or, of course, the printed words, has to tell the examiner whether the word pairs are the same or different. The Wepman is designed for younger students, but it has some useful discriminative value for adults too. One student, however, who came in during the early days of Arrowsmith necessitated a new test.

Seventeen-year-old Dana Holmes hesitated a great deal on the Wepman test but did not make a single error. Clearly, though, she had an auditory speech discrimination disorder. Dana described how hard it was for her to listen in class (for example, she heard *vulture* as *fulture*), but I could not measure this on the Wepman test. She dropped out of her last year of high school because she found it too stressful and embarrassing to be always asking her teachers to repeat words used in class. Dana completed high school through correspondence courses to avoid having to listen to lectures. For her, I developed a more challenging auditory speech discrimination test using forty pairs of multisyllabic words with differences embedded within (*demote/denote, disallow/disavow*). On this test, which we now use widely with adults, she made sixteen errors.

Over the many years we have administered the Arrowsmith Auditory Speech Discrimination Test, we have observed that those who make eight or more errors on the test generally share a common trait: they have dropped French (or other second-language courses) before eighth grade because they had difficulty discriminating the speech sounds.

To experience what it feels like to live with this deficit, place some tissue in both ears. Use only enough to lightly block the sound from your surroundings. Now turn on the radio or have a conversation with someone on the telephone. (You can achieve the same effect without tissue. Turn on the radio and set the dial so it's just off the frequency for your favorite station—ideally one with lots of talk.) Is it hard to clearly hear what's being said? Do you have to pay extra attention? Does the other person's voice sound fuzzy or as if he or she were speaking a language other than English? Do you get frustrated or fatigued with the effort? Do you tune out or cut short the conversation? This is the experience of a person with an auditory speech discrimination problem.

In his book *Higher Cortical Functions in Man,* Luria describes how the brain processes speech. He knew that for humans to articulate or comprehend speech, they must identify and amplify what he called "distinguishing phonemic signs" (the speech sounds that make up words) from other nonspeech sounds in the environment, such as, say, the rustling of leaves. This process starts in infancy. When this ability to make those distinctions is compromised, said Luria, "the boundary between hearing speech and understanding it loses its sharp distinction." Such a person can hear speech but can't make sense of it.

Someone with this deficit will typically ask another person addressing her to slow down or to enunciate. Reading lips or trying to guess the words using context are other common strategies. But clearly, given a significant level of difficulty, those strategies are neither effective nor often available. To have this deficit is to always be straining to understand spoken language, and the result is extreme fatigue. The more fatigued you become, the more pronounced the inability to process language.

While taking notes in a classroom, say, you mishear some words and later discover that your notes make no sense. You have trouble deciphering the country of origin of someone who speaks English with an accent, and you have trouble understanding accented English, so that an American with this deficit watching a British film may be grateful for subtitles.

If the deficit is serious, speech can sound fuzzy—like white noise or as if people were speaking a foreign language. Spelling is affected, as is speech pronunciation, because you spell and say the word as you (incorrectly) hear it. Learning the sounds of letters in the early stage of reading is also affected, as described in Chapter 13. When the speech signal is degraded acoustically by background noise (in a noisy room), speech discrimination and therefore comprehension suffer considerably more.

In Ann Tulloch's case, the word *them* sounded to her like *vem*—which of course makes no sense. It's a bit like trying to rely on high school French to comprehend the rapid-fire language of a francophone. Depending on your facility and the speaker's speed, you hear a sentence but mishear every third or fourth word. Maybe critical words. Out the window goes comprehension. In a very real and functional way, it is as if you are partially deaf.

The telephone holds a particular terror for those with an auditory speech discrimination problem because they cannot see the speaker's face and attempt to read lips to compensate for the difficulty. And that was certainly the case with Ann.

"The phone would ring at work," Ann told me, "and my boss would say, 'Ann, go get it.' And I'd be sweating. Before doing the auditory speech discrimination cognitive exercise, I couldn't recognize whose voice it was on the phone—even someone I knew. I felt so stupid. I couldn't tell if the person was a man or woman. I just couldn't tell the voice. And after working on this area, now I know."

Ann came to us in her mid-thirties, and by then she had developed certain strategies. Sometimes they worked, but often they didn't.

"I've always gone through life second-guessing people," she explained. "I couldn't hear properly what they were saying, and I couldn't remember [owing to another deficit—a memory-for-information problem]. I'd fake it. I'd size up what was going on and then try to work my way around it so I didn't appear stupid or as though I didn't know what was going on. But half the time I didn't."

One casualty of this brain deficit is the ability to enjoy a song, because the words go missing too. Ann told me that unless a word was well pronounced or part of a chorus, the lyrics were indecipherable. And when the impairment in her brain was addressed, songs had new clarity for her. Here she is describing that moment: "The experience of being able to hear lyrics in songs was so powerful and so sudden that to this day I remember exactly where I was in my car—in Toronto on the Rosedale Valley Road approaching the Bayview Extension to the Don Valley Parkway. I started to be able to hear the different words in songs. More pronounced. I could understand each individual word. Before, everything just flowed together for me in songs. Before, it was like noise."

What led to this change for Ann? The cognitive exercise for this deficit involves listening to speech sounds in unfamiliar languages (Swahili, Kurdish, and Bengali, to name a few). An English speaker finds it challenging to discriminate one such sound from another, placing a load on the brain's level of activation beyond the normal demands required to hear familiar sounds. But over time and after repeated exposure, Ann's ability to discriminate speech was enhanced.

When I interviewed Ann in March 2011, she talked about a job she had held for seven years (2002 to 2009) at General Motors. She knows unequivocally that she could not have done the job without addressing her auditory speech discrimination problem. The job involved the pricing and invoicing of automotive parts, which meant talking on the telephone to car dealers all over the United States. Ann had to understand people with all sorts of accents, some who spoke rapidly, some who spoke softly. She had to accurately hear what they needed in order to respond to their requests, and she had to quickly recognize the different callers by their voices. Ann was so good at the job that she took on all the difficult requests, and dealers would specifically ask for her. No longer hampered by her deficit, she could excel and was rated the top employee in her department.

You met little Zachary earlier. He was the boy in Toronto who could make no sense of his world. "What what what?" and "Huh huh huh?" were the words he used most. Among his several deficits was auditory speech discrimination, which played out in his case like this: Zachary's father told him that they were going to a pet store called Big Al's. "We're going to Big Al's tent sale," his father announced, which greatly excited Zachary because he heard "Big Owl ten sale" and he was later very miffed to see not one owl, let alone ten. This led to a tantrum because, as far as he could see, his father had misled him.

Once at a party, he asked someone to pour him a glass of Pepsi. But what he said was "sexy." If you said "Pepsi" to him, he heard "sexy," and he therefore learned the word incorrectly. He would play the card game Go Fish, but Zachary called it Goldfish—and that's what his family came to call it. In school, he got into trouble saying he didn't want to do "arts and crap" (he meant to say "arts and crafts" but he had misheard and learned the term as "arts and crap" and that's how he, in all innocence, repeated it). As a reasult, he was viewed as a troublemaker.

The interviews for this book were recorded and then transcribed by an agency. My heart went out to one particular transcriber who clearly has an auditory speech discrimination deficit. Some of the errors that quickly identified her deficit were these: "She wasn't cold and awful and evil," became, "She wasn't cold and off the needle." Elsewhere, "eye patches" became "high taxes," and "Catskills" (as in the mountains) became "cats' gills." This woman, I am sure, suffers in her daily life. But her burden is that much greater because she's chosen a job that requires constant speech discrimination.

CHAPTER TWENTY-TWO

THE IMPACT OF LEARNING DISABILITIES

Some Arrowsmith students at the American Christian School in Suc-casunna, New Jersey, were playing outside and chanced upon a cicada that was spinning on its back and delivering its signature elec-tric song—but in this case, one of distress.

"The kids wanted me to go out there and see this," said Carol Midkiff, the principal. "The cicada was flipped over on its back. So I got a pencil and turned it over. It tried to fly, but one wing was a little higher than the other, and it couldn't. I put it in an envelope and brought it over to the hedgerow. I felt such pity for this cicada. Nothing that it did caused this."

For Midkiff, the cicada was a metaphor for children who cannot ver-balize their pain and who will never take flight. She was thankful that her grandson, who was among those students who found the insect that day, had been spared that fate.

When I went into Toronto Arrowsmith classrooms in October 2010 and asked students how learning challenges had affected them, I was reminded of my own experiences.

When given the opportunity to share their feelings about their strug-gles, students did not hesitate. Their honesty and openness were powerful, and no one, students or staff, was left untouched by what they said. Every boy and girl, every young man and woman sitting in those classrooms, looked like students in any classroom in the world—a mix of black hair, blond hair, blue eyes, and brown eyes. If not for the little bit of probing that day, I would never have known the individual pain and suffering that each had endured as a result of learning disabilities.

Speaking with these students reminded me how I too had worked to hide the pain from everyone around me, including my family and friends. The students had made, as I had once made, a personal pact of silence.

In the younger and intermediate grades (ages six to fifteen), the tissue box made its rounds. In the young adult classroom (students aged eighteen to thirty), there were fewer tears, but the stories were every bit as raw and heartbreaking. In both groups, a student revealed that during his darkest days of struggle at school, he had thought seriously, as I had, that suicide was the only option. One student described how, at the age of fifteen, she started to cut her own skin—"symbolic suicide," she called it. "Everyone goes to school to learn things, I go to school to feel stupid," she said.

In the intermediate class, a boy named Andrew described what it was like to shoulder two intricately linked burdens: profound learning challenges and daily bullying. "I'll never forget this," he said. "These kids had a gang, and whenever I went to the bathroom, they'd follow me, and they just called me stupid. And one day, they started with the physical stuff, and they beat me up every single day. And they warned me if I told anyone, they'd come after me. Day after day, I'd get punched and kicked, and I just wanted to not go to that school anymore. I wanted to die. I wanted it all to end."

"Do you think you were targeted because you are learning disabled?" I asked him.

"Yeah," he replied. "And the funny thing is, one guy who beat me up has a learning disability."

Jason Grant, a young adult, described what it was like to be capable and incapable at the same time. "With projects in class, there were certain things I could not do at all. And there were things where the teachers were utterly shocked at how good my work was." Jason gave the example of an illustration he had drawn as a boy after reading a book by British author Ted Hughes, *The Iron Man: A Children's Story in Five Nights*:

The teacher was utterly blown away by what I drew. He said, "This is the most thinking I've ever seen a grade-three pupil put into a picture." And yet there were other things that I'd struggle with a lot. So I was really confused. How come I can do that so well, and how come I can't do this? After so many years, it hurt a lot, and I got into this really dark rut. Throughout the second half of high school, I said to myself, "If things don't shape up in the next year or two, I'm done." I was so ready. I even had a date set. This is actually the first time I've told anybody this. I just felt like there was no end to the suffering.

This is how it is to have learning disabilities. You excel at tasks that call on your strengths, and you struggle or fail at tasks that call on your weaknesses. This is a confusing world for the person living it and equally confusing for teachers and parents who wonder, "If he can do so well here, why can't he do well over there?" Thus the comments, "If you just tried harder, you could do better" or "You just must not be motivated."

The student comes to internalize this view. *They say I'm lazy and unmotivated; it must be so.* And the individual learns to fear high performance lest it be held up as a general standard—one impossible to maintain when specific learning problems are in play.

Students in the senior class described being caught in a bind. Some people, they said, mistakenly believe that those with learning disabilities are stupid. And if you're not stupid, the circular argument goes, well then you can't possibly have a learning disability. Two students said that their siblings accused them of faking learning challenges to get attention. One young woman told us that her husband refused for years to believe she had a learning disability until he saw her spelling in an essay. Everyone in that class had borne one or more of these labels from parents and teachers: lazy, difficult, obstreperous, messy, disorganized, unmotivated. Or they had been told, "You have so much potential if you'd only apply yourself." One student was informed by a teacher that he would never graduate from high school. Another student described being told by a college professor that he was "garbage and without talent." He dropped out, saying he had too much self-respect to take more insults like that. Yet another remembered that from primary school to college, he was never given credit for his efforts.

There are compassionate teachers everywhere, and I heard stories that day of teachers helping: buying a student a tape recorder to help with note taking, staying after class to review material, purchasing journals to help students organize themselves. But the "garbage" remark provoked an outpouring of similar stories from these students, suggesting that a great deal needs to be done to educate all teachers about learning disabilities. And yet Tara Anchel, the assistant director of Arrowsmith School who accompanied me that day, remembers getting just a single one-hour lecture on special education (with learning disabilities only a part of it) during her entire time in teachers' college.

Clear from the testimony of these students was that thoughtless remarks by teachers, who typically meant no harm, made a permanent impression.

A child without learning disabilities might shrug off such a comment, but it can go straight to the heart of a student with a history of failure and disappointment. As educators, we need to be aware of these students' carefully hidden vulnerabilities.

As a culture, we do not provide an environment where students can openly discuss their struggles with their teachers or enlist a teacher as an ally. Listening to all the students, parents, and teachers I interviewed for this book, I was struck anew by the fact that there still is a stigma attached to having a learning disability. This fosters a climate of shame, so that a child works hard to hide learning problems from both parents and teachers. As long as we keep this hidden in the shadows, behavior will be wrongly attributed to willfulness or laziness or some other erroneous explanation rather than to a real incapacity as a result of a learning disability. I experienced the same thing fifty years ago.

Many students that day talked about feeling guilty because they felt that they were a source of disappointment to their parents or because their learning disabilities had caused their parents anguish. Virtually every student had been traumatized by failure in school, and the damage done there had carried over into home life and social life.

One student remembered what he called "major test anxiety," which he still feels to this day. He recalled an incident in eighth grade when he walked into class for a history test and his mind went blank. "My anxiety," he said, "just went through the roof to the extent where I actually started to break down and cry in the middle of this test. So I'm sitting there, I'm thirteen years old, everyone is standing around me, and I'm crying because I can't read the question or figure out what it's asking of me." His learning disability had made it hard for him to read and comprehend the test; his anxiety had robbed him of whatever cognitive functioning he then possessed. He was later teased mercilessly.

Performance anxiety is very common among those who are learning disabled. Initially confined to their learning weakness, the apprehension soon affects performance even in areas where the student has strengths. The student may become anxious in any learning situation, and in time anxiety becomes its own separate problem, shutting down all learning.

I have seen many adults who have spent their lives coping with, covering up, or compensating for their learning problems and the subsequent emotional impact. These men and women create codified defense systems comprising various strategies: creating rigid rule structures, compulsive list making, avoiding new situations, procrastination, perfectionism, and deflecting attention, to name a few.

Ten years ago, I received a phone call from a person who would not identify himself other than to say he was a successful entrepreneur who had struggled with writing all his life. He described the elaborate tactics he had devised to hide this problem. On one occasion, when he had to write an important exam, he showed up with a cast on his writing hand so he could do the exam orally. He had developed, he said, a brusque and aggressive persona at work so his employees would keep their distance and not discover his difficulty. If asked to write, he would bark, "That's what I have a secretary for!" Only two people knew of his problem: his wife and his closest friend. His wish was that others would not have to hide as he had.

At Arrowsmith, we openly talk about how the different cognitive problems have affected these students' lives. And when new strengths emerge as these problems are addressed, we discuss the changes. Each student's experience is recognized and validated.

The earlier we can begin to address these challenges, the better. Early assessment and treatment allow young people to make career choices based on a wider range of abilities.

One floor above Andrew's classroom, in my office at the Toronto Arrowsmith School, hangs a beautiful piece of framed modern art in earthy shades of brown, with a Chinese ideogram that spells the word *courage*. I bought the piece to remind myself of what it takes for students to face their learning challenges and to do the work to overcome them.

For a long time, the statistic most commonly cited to measure the incidence of learning disorders in North America was one in ten. Some more recent surveys have suggested the figure is lower; I believe the true incidence is at least 10 percent and could be as high as 20 percent.

Why might surveys be reporting lower figures? For one thing, the term *learning disabled* is a familiar one, but many people are convinced, mistakenly, that it doesn't describe them or their children. Parents may know that their child is experiencing some difficulty in learning, and they want the child to be tested, but perhaps that has yet to be arranged. Some parents are reluctant to reveal their child's learning disorder to anyone. Finally, many who are diagnosed with a learning disability may later drop out of school and stop reporting themselves as learning disabled.

I interviewed one parent of a child at Arrowsmith, a public health scientist who researches social issues. She argued that learning disabilities could be seen as a public health issue. "LD is more hidden somehow, more invisible, and people tend to be blamed for their problems," she said. "There is a stigma associated with learning disabilities. If you have severe learning disabilities like my son, your mental health is going to be affected." She pointed to studies showing that those on the margins of what is considered normal are more likely to indulge in risk-taking behavior. And she wondered if such behavior is connected to the bewilderment that those who are learning disabled feel. A failure to fathom why they can't do what others appear to do easily can lead to a fatalistic attitude. They ask themselves, "Why should I try? What does it matter? I don't have a good future anyway."

Even someone as informed and aware as this public health scientist concedes that she delayed admitting to and coping with her son's difficulties. "I didn't immediately say, 'You have learning disabilities.' I think I should have told him right away, 'I know you are feeling bad and we are going to do everything we can to get you the help you need.' Because when I finally did say to him, 'You learn differently from other kids,' it was a huge relief for him. So I wish I had done that sooner."

I hear this over and over again—this notion that parents are protecting their children by not telling them they have learning problems. Some parents believe that the problem cannot be fixed; some believe in focusing on their children's strengths. But a child with learning disabilities knows something is wrong. Living without an explanation for the problem is more stressful than coming to terms with it. Until the stigma associated with self-disclosure fades, we will continue to get low figures in these statistical surveys. The testimony of individuals in this book offers a grim reminder of the shame associated with learning disorders, the stress it adds to families and marriages, and the terrible suffering endured all around. The prisons teem with men and women who struggled to learn as children, and again the numbers vary, from 38 percent to 80 percent. Whatever the true numbers of learning-disabled inmates, they are unbearably high.

How do you measure the cost when a job is lost because of a learning disorder? Or when a marriage buckles under the strain—emotional, financial—because one or both spouses are learning disabled? How many suicides are linked to an inability to learn? It's impossible to say.

On a grander scale, the cost to society has been estimated. A world summit on learning disabilities held at Lake Louise, Alberta, in 2008 reported that learning disorders in Canada mean $33 billion a year in lost productivity. Related health care costs added between $10 and $20 billion to that figure. American figures, given the difference in population size, are likely on the order of ten times those numbers. Calculating the true cost of learning disorders is impossible. This I do know: the pain associated with learning disorders is both excruciating and avoidable.

Norman Doidge began to see that some of the emotional issues that he dealt with in his psychiatric practice were the direct result of learning disabilities:

I saw people getting better cognitively with the Arrowsmith exercises, and I saw children, young people who had always desperately wanted to learn, suddenly able to become avid readers and enjoy the learning that they always sensed they longed for. I also saw, as I was never able to see before, the absolute devastation, emotionally, that their learning disorders had caused them. Why could I not see it before? Because they were so trapped in the present, in the talking about the unfinished essay, trapped in avoidant behavior, trapped in oppositional behavior. All of which I began to realize weren't primary, but were secondary defense mechanisms covering over their cognitive tragedies.

I also saw that as they improved cognitively, the emotional damage of the learning disorders didn't go away immediately. They enjoyed the present, but their sense of confidence was often still haunted by years of thinking of themselves as stupid or lazy. And so, ironically, it was back to doing psychotherapeutic work—to help them understand what they had been through. Once they were liberated from their cognitive deficits, they could go back and deal with other garden-variety psychotherapeutic issues and make progress at times much more quickly.

It still makes me very sad to think about the effects of these learning disorders. And this is a side effect of plasticity. This is a kind of overlearned behavior (the negative view of self resulting from one's experiences living with the cognitive deficits), plastically wired into their brains that's not so easy to get rid of, which is why anyone who wants to think this question through with any level of profundity, realizes that plasticity isn't simply our friend. It's a property of the human brain and it can give rise to things that we like or loathe.

A child with a learning disability can be told innumerable times that she is competent, but who can blame her for not believing it when her own experience flies in the face of that encouragement? This is why I feel so strongly about addressing these cognitive problems as early as possible, so that a person can have a positive experience as a learner and develop a healthy self-concept based on real competence.

The stigma associated with having a learning disability will ease when we all understand that we can accomplish what was once thought impossible: we can change, fundamentally and profoundly, our capacity to learn.

CHAPTER TWENTY-THREE

WORD SPREADS

2001

A turning point in my life's work came on May 26, when a national Canadian magazine, *Saturday Night*, ran a feature story, "Building a Better Brain," by Norman Doidge. A patient's mention of our program brought him to Arrowsmith. Out of curiosity, he came to the school, tried some of the brain exercises, and started interviewing our students and their parents. Then he began referring some of his own patients to us. Ours was the first practical, clinical example he had seen of neuroplasticity at work.

"The results," Doidge said, "were often amazing. I saw the progress of the people I had referred, and I knew it was real."

His experience at Arrowsmith—"an extremely important encounter," he calls it—changed him as a clinician and as a human being dealing with other human beings. Doidge began to see that until he understood someone's cognitive strengths and weaknesses, unique brain makeup, and wiring, it would be hard for him to understand that individual's worldview, or "map of the world," as he calls it.

Published a year after our first encounter, the *Saturday Night* piece—which ranged over neuroplasticity, Luria, Zazetsky, my childhood, and Arrowsmith—was informative and poignant.

The effect of the piece was immediate and far reaching. Enrollment took off. We found—because we had to—a much bigger space for the school: a three-story rundown building on St. Clair Avenue. We gutted it, made it new, and moved in. Then we did the same with the building next door. The magazine piece also stirred interest among people in Vancouver and led to the founding of a school implementing the Arrowsmith Program there.

Today, there are thirty-five Arrowsmith Programs all over Canada and the United States, with several more in the works. Every summer, about thirty new teachers arrive at the lab school in Toronto and take the three-week course that will enable them to become Arrowsmith teachers in their home states and provinces.

Every August, I feel humbled when I meet these teachers—all with a great passion for making a difference in the lives of learning-disabled children. All are frustrated by the status quo; all are open to new approaches to learning disabilities.

My vision is of a world in which no child ever struggles with a learning disability, no child is ever stigmatized as having one, and no child experiences the ongoing emotional pain of living with a learning disability.

My vision is for all schools to become places where children can go to strengthen their brains so they can learn effectively and efficiently. Cognitive exercises, using the principles of neuroplasticity, will become an integral part of each school's curriculum. In this way, learning problems can be addressed early, as part of a regular curriculum, and students without cognitive deficits will all benefit from cognitive stimulation.

In the interim, my goal is that every child be assessed at an early age, their brain deficits (major or minor) clearly determined, and tailor-made exercises applied to overcome any learning problems. In this way, with early intervention, no negative patterns of behavior will get entrenched.

That people with learning disabilities don't dare to dream breaks my heart. We now have the tools to address these problems, strengthen and rewire and improve their brains, and avoid a tremendous amount of needless suffering.

I am passionate about this work, about its ability to change and improve lives. My daily prayer is that this work, grounded in compassion, its integrity uncompromised, goes out into the world with ease and grace.

DESCRIPTION OF THE COGNITIVE DEFICITS
ADDRESSED BY THE ARROWSMITH PROGRAM

This appendix lists nineteen cognitive deficits, along with descriptors of the common features and a phrase in italics typically used in connection with each one. Someone with that particular deficit is likely to utter that phrase, sometimes in an effort to explain the deficit away. There wasn't space in the book to discuss every deficit in full; I focused on the ones that have the greatest impact on people's lives.

The Arrowsmith cognitive assessment identifies which cognitive areas are underperforming and contributing to an individual's learning difficulties. The test results serve as the basis for developing an individualized program of cognitive exercises for each student.

As you read the descriptors, keep in mind that the list is neither exclusive nor exhaustive. The intricate and constant interplay of cognitive areas makes it difficult to offer a fixed list of symptoms for any one cognitive area.

Some of the same symptoms may be listed in more than one cognitive area. This is because those particular indicators (inability to plan, for example) may occur as a result of a deficit in each of those areas (symbolic thinking, artifactual thinking, predicative speech, for example); the manifestations can look the same to the observer but have a different underlying cause.

Finally, if the common features of a deficit in a particular cognitive area don't apply, an individual may have a strength in this area.

Motor Symbol Sequencing

"Please don't erase that blackboard yet."

This capacity, centered in the premotor region of the left hemisphere of the brain, is involved in the process of learning motor plans necessary to consistently and sequentially produce a set of symbols (the alphabet, for example, or numbers). When there is a weakness in this capacity, processes involving input through the eye (reading) and output through the hand (writing) and mouth (speaking) are impaired to varying degrees.

This deficit means the eyes can't track properly, so that misreading becomes an issue. (You read *step hall* for a road sign that says *steep hill.*) Your handwriting is messy, irregular, and not automatic. You must focus so much on writing itself that copying material (from, say, blackboard to notebook) is difficult. Your spelling is erratic, your speech is rambling. In telling a story, you leave out chunks of critical information, making it hard for others to follow you.

Symbol Relations

"I just don't get it."

This capacity is involved in understanding the relationships between two or more ideas or concepts. The deficit is centered in the juncture of the occipital-parietal-temporal region in the left hemisphere of the brain. In more severe cases, you reverse letters, such as *b* and *d* or *p* and *q,* long after this is developmentally appropriate. You have trouble learning how to read an analogue clock.

You can learn math procedures but not the why of the procedure. Math and logic are often about relationships (percentages, fractions) that you cannot fathom. When the weakness is severe, you have trouble understanding cause-and-effect relationships or why events happen. This has implications for learning in school, on the job, and in social situations.

Prepositions (*with, without, in, out*) are also about relationships, so they too are hard to understand, as is grammar. Despite reading material repeatedly, you are never certain you have understood.

Memory for Information or Instructions

"I have a memory like a sieve."

This is the capacity for remembering chunks of information. You have trouble remembering and therefore following lectures or extended conversations or instructions. Instructions have to be repeated several times before you retain them. This deficit, like the others named so far, involves the left hemisphere of the brain—the temporal region in particular.

Following a radio program or newscast can prove difficult. You tend to hang back and not participate in conversation. You tune out conversations because so much effort is required to retain the information.

Predicative Speech

"My words don't always come out in the right order."

With this deficit, the neurological process that converts thought into an organized sequence of words is flawed. You cannot learn the rules governing sentence structure. You speak and write in short sentences and have a hard time following long sentences. As compensation, you may keep a store of memorized short phrases. And worse, because you can't mentally rehearse through internal speech what you're going to say or do, you cannot anticipate the consequences of your words and actions and may appear to be rude and lacking in tact.

Oral, written, and inner speech expression are all affected.

Broca's Speech Pronunciation

"People say I mumble."

People who struggle to pronounce words have a weakness in the frontal lobe area in the left hemisphere of the brain known as Broca's area, named after the nineteenth-century French anatomist Pierre Paul Broca, who first discovered the link.

You mispronounce words or avoid using words you know and understand because you're uncertain about pronunciation. This may restrict spoken vocabulary to simpler words.

Since speaking requires concentration, it's hard for you to talk and think at the same time, and you easily lose your train of thought. Unless you work from a prepared text, public speaking is hard. At a severe level, your speech tends to be flat and monotone, with a lack of rhythm and intonation. There is a tendency to mumble. This impairment also interferes with the ability to learn a foreign language. Additionally, when learning to read, you have difficulty converting letters into sounds.

Auditory Speech Discrimination

"I'm sorry. Could you repeat that?"

This is the capacity to discriminate between similar-sounding speech sounds (*fear/hear, doom/tomb*). Part of the superior temporal region in the left hemisphere of the brain is affected.

Symbolic Thinking

"Planning was never my strong suit."

This is the capacity for mental initiative. The left side of the brain (specifically the prefrontal cortex) is the one implicated.

You have great difficulty developing strategies for studying. If shown a study method, you may be able to follow it, but you cannot develop your own. You are easily distracted and frequently called to task for having a short attention span. Organization, planning, self-direction, and establishing long-term goals are all major challenges. As a result, you live for the moment, and others may view you as untrustworthy or flighty.

Symbol Recognition

"I was never a great reader."

The part of the brain not working here is in the left hemisphere, more specifically the left occipito-temporal region, which allows us to recognize and remember a word or symbol.

You have to study a word many more times than average before you can memorize it, and thus recognize it and say it correctly the next time you see it. In many cases, you cannot learn sight words even with multiple repetition, and every time you are presented with a word that should be familiar, you need to sound it out as if you are seeing it for the first time. As a result, learning to read and spell is a slow process.

Lexical Memory

"I'm not good at remembering the names of things."

This is the capacity for remembering words. You have trouble remembering individual words and the names of things—the days of the week, the names of colors and the names of people. The temporal region in the left hemisphere of the brain is the one involved.

Kinesthetic Perception

"I am such a klutz."

This is the capacity to perceive where either side or both sides of your body are in space. The part of the brain not fully engaged is the somatosensory area in the parietal lobe in either the left or the right hemisphere, and sometimes both. You have a tendency to bump into objects with the affected side of the body. Driving a car or using power tools can be risky. If the problem occurs in your writing hand, there is uneven pressure, and you may wander off the line.

Kinesthetic Speech

"I slur my words sometimes."

In this deficit, there is a lack of awareness of the position of the lips and tongue, resulting in slurred speech. This is very much a kinesthetic perception problem, but in this case, what's affected is the specific area in the brain that controls feedback to the tongue, lips, and mouth for clear articulation of speech. You have trouble rapidly repeating a tongue twister such as "truly rural" or "three free throws." The left or right hemispheres of the brain might be involved, or both.

Artifactual Thinking

"I'm just not good at reading people."

This capacity is necessary for you to interpret emotions and modify your behavior accordingly. The right side of the brain, and specifically the prefrontal cortex, is the one implicated.

You cannot interpret nonverbal cues and information such as facial expressions and body language, and as a result you can't modify your behavior according to the signals people send you. Unable to "read" nonverbal information coming from your boss, your teacher, or your friends and relations, you don't always act appropriately and you can't self-correct. There is a failure to understand others coupled with a failure to understand yourself.

Narrow Visual Span

"My eyes hurt when I read."

This is the capacity responsible for the number of symbols or objects one can see in a single visual fixation. The occipital lobe—in the left, right, or both hemispheres at the back of the brain—is the problem area. When the span is restricted, you cannot see whole words in a single visual fixation. You must make three to ten times the normal number of eye fixations to read a line of printed material. You suffer easily from eye fatigue, and your reading is slow.

Object Recognition

"Have we met?"

This is the capacity for recognizing and remembering the details of visual objects, including faces. This deficit is centered in a network of right hemisphere areas. You take longer to recognize and locate objects—when shopping or looking for something in the refrigerator or landmarks in your own neighborhood. You have trouble recognizing and remembering faces and miss details in facial expressions, which creates social and interpersonal problems for you.

Spatial Reasoning

"I am forever getting lost."

Spatial reasoning, linked primarily to the right parietal area of the brain, is the capacity to imagine a series of moves through space before executing them. When this capacity is weak, you cannot map out inside your head how to get from one place to another or rotate the map inside your head. You forget where you've left objects because you aren't able to create a spatial map of those objects and their location. Your work space tends to feature material stacked in piles, within sight. If you put something away in a filing cabinet or drawer, you later have trouble imagining in your head where it is.

Mechanical Reasoning

"I'm not handy."

A mechanical reasoning problem, centered in the right hemisphere, means you have difficulty imagining how machines operate and how their parts interact with one another. You also have trouble handling tools effectively.

Abstract Reasoning

"I couldn't program the VCR to save my life."

Some tasks require an internal sequential logic, and the order of procedure is paramount. A person with an abstract reasoning problem, also a right-hemisphere issue, has trouble carrying out in sequence a series of steps in a non-language-related task, such as navigating a computer program, preparing a complex recipe, or programming the VCR. Computer programmers are often good chefs; they know about the proper order of things. With this deficit, you don't.

Primary Motor

"My reaction time is a bit slow."

This problem interferes with the speed, strength, and control of muscle movements on one side of the body or the other. This results in awkward body movement and less-articulated movement on the affected side of the body. The primary motor "strip," as it's called, behind the prefrontal cortex, is the region of the brain affected. The left, right, or both sides may be affected.

Unlike kinesthetic perception, which provides feedback to guide and modify movement (think of a receiver in football running an intricate pass pattern), primary motor simply tells your muscles to move (enabling that same receiver to almost reflexively catch the ball). And as with so many parts of the brain, kinesthetic perception and primary motor work together.

Supplementary Motor/Quantification

"I'm not a numbers person."

A problem in this area means you can't do math in your head. Simple counting, calculating change, learning to add, and the multiplication tables can all be problematic. The deficit is centered in an area in the parietal lobes related to understanding quantity and number.

Human behavior is complex, and no single factor can explain it all. We can be affected by psychological trauma, our cultural background, our personalities, our upbringing. But these nineteen cognitive areas, though clearly not an exhaustive list of all the areas in the brain related to learning, are key pieces. A strength or weakness critically shapes how we participate in the world.

Many students come to an Arrowsmith Program with half a dozen or more deficits, some of them rated severe. Individually, these deficits can be a great burden on those who have them, and more so when deficits coincide and conspire with one another.

LOBES OF THE BRAIN

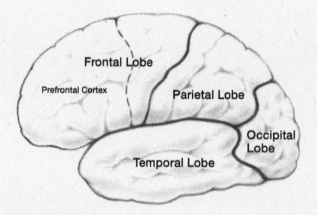

Figure 9. Lobes of the Brain.
This is a side or lateral view of the human brain showing the four major lobes:
Frontal—executive functions, thinking, planning (especially the prefrontal cortex), motor planning and execution
Parietal—perception of body sensations, spatial perception
Occipital—visual perception and processing
Temporal—auditory perception and processing, memory

BRODMANN AREAS OF THE BRAIN

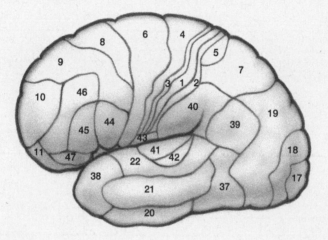

Figure 10. Brodmann Areas.

Korbinian Brodmann (1868–1918) divided the human brain into 43 to 47 distinct regions based on cellular composition and organization. These areas are still referred to in current neuroscience literature. We now know, through imaging studies, that there are multiple subregions within many of these areas, with different functions. This image is a lateral view of the brain, so areas that appear on the medial surface of the brain are not represented in this view.

The Brodmann areas corresponding to the cognitive areas described in the book are outlined in the chart below. Within many of these areas there are multiple subregions that have not been noted on the illustration or in the chart. Also, no area works in isolation, multiple areas combine in neural networks to contribute to the processes described in this book.

Brodmann Areas Referred to in Cognitive Chapters

Chapter	Brodmann Area	Area Name	Arrowsmith Cognitive Area
Chapter 8 Lost in Translation	39	Angular Gyrus—juncture of parietal-occipital-temporal regions	Symbol relations
Chapter 9 Hitting the Wall	8, 9, 10, 11, 44, 45, 46, 47	Left Prefrontal Cortex	Symbolic thinking
Chapter 11 Leap Before You Look	8, 9, 10, 11, 44, 45, 46, 47	Right Prefrontal Cortex	Artifactual thinking
Chapter 12 When a Picture Does Not Paint a Thousand Words (See Figure 3 in Chapter 12 for areas involved in face recognition)	37 18, 19	Right Hemisphere Fusiform Gyrus in Occipito-Temporal Region Occipital Areas—secondary and associative visual processing	Object and facial recognition
Chapter 13 A Closed Book (See Figure 4 in Chapter 13 for a map of other areas involved in reading)	37 18, 19	Left Hemisphere Fusiform Gyrus in Occipito-Temporal Region Occipital Areas—secondary and associative visual processing	Symbol recognition (visual word form area/ brain's letterbox)
Chapter 14 Nothing to Write Home About	6	Premotor Region	Motor symbol sequencing
Chapter 15 Blind to One's Own Body	3, 2, 1 4	Somatosensory Cortex Primary Motor Cortex	Kinesthetic perception Primary motor
Chapter 17 Lost in Space	5, 7	Parietal Lobe	Spatial relations
Chapter 20 When 2+2 Does Not Equal 4	Sections of Area 7, Area 39, Sections of Area 6	Horizontal Intraparietal Sulcus Angular Gyrus Supplementary Motor Area	Quantification
Chapter 21 In One Ear and Out the Other	22	Superior Temporal Region	Auditory speech discrimination

NOTES

INTRODUCTION

1   Zazetsky's next memory: A. R. Luria, *The Man with a Shattered World*, trans. Lynn Solotaroff (Cambridge, MA: Harvard University Press, 1972), 8–9.

1   "Since I was wounded": Ibid., 135.

2   "Before I've had a chance": Ibid., 154.

2   "I'm in a kind of fog": Ibid., 11–12.

2   "My memory's a blank'": Ibid.

3   "This strange illness I have": Ibid., 145.

3   "Precise knowledge was rarely to be found": Ibid., 23.

4   "His [Zazetsky's] description is exceptionally": Ibid., 86.

5   The phrase *learning disabled* was coined: Samuel A. Kirk and Barbara Bateman, "Diagnosis and Remediation of Learning Disabilities," *Exceptional Children* 29 (1962): 73.

6   "I'm in a kind of fog": Luria, *Man with a Shattered World*, 11–12.

6   "The bullet that penetrated this patient's brain": Ibid., 113.

7   "I knew what the words 'mother' and 'daughter' meant": Ibid., 132.

7   During this time, I came across the research: Mark R. Rosenzweig, D. Krech, E. L. Bennett, and M. C. Diamond, "Effects of Environmental Complexity and Training on Brain Chemistry and Anatomy: A Replication and Extension," *Journal of Comparative and Physiological Psychology* 55 (1962): 429–437. Mark R. Rosenzweig, "Environmental Complexity, Cerebral Change, and Behavior," *American Psychologist* 21 (1966): 321–332. Mark R. Rosenzweig, W. Love, and E. L. Bennett, "Effects of a Few Hours a Day of Enriched Experience on Brain Chemistry and Brain Weights," *Physiology and Behaviour* 3 (1968): 819–825.

CHAPTER ONE: THE ANATOMY OF RESISTANCE

10   Norman Doidge, the author: Norman Doidge, *The Brain that Changes Itself* (New York: Penguin Books, 2007).

10   Every field of science has foundational beliefs: Thomas S. Kuhn, *The Structure of Scientific Revolutions* (Chicago: University of Chicago Press, 1996), 5.

11   In an article published in fall 2010: Katie Ronstadt and Paul B. Yellin, "Linking Mind, Brain and Education to Clinical Practice: A Proposal for Transdisciplinary Collaboration," *Mind, Brain, and Education* 4 (2010): 95–101.

12   Only since 1990: P. N. Tandon, "The Decade of the Brain," *Neurology India* 48:3 (2000): 199–207.

12   I vividly remember: Gerd Kempermann and Fred H. Gage, "New Nerve Cells for
      the Adult Brain," *Scientific American* (May 1999): 48–53.

12   Only as recently as 2000 did Eric Kandel: Eric Kandel, "Nobel Lecture: The
      Molecular Biology of Memory Storage: A Dialog between Genes and Synapses,"
      presented at Karolinska Institute, Stockholm, December 8, 2000. http://www.
      nobelprize.org/nobel_prizes/medicine/laureates/2000/kandel-lecture.html.

12   "Consider the possibility": Santiago Ramón y Cajal, *Advice for a Young Investiga-
      tor,* trans. Neely Swanson and Larry W. Swanson (Cambridge, MA: MIT Press,
      1999), xvi.

13   In 1783, an Italian anatomist named Michele Vincenzo Malacarne: Celine Che-
      rici, "Vincenzo Malacarne (1744–1816): A Researcher in Neurophysiology
      between Anatomophysiology and Electrical Physiology of the Human Brain,"
      Comptes Rendus Biologie, 329 (5–6) (2006): 319–329.

13   Jerzy Konorski, a Polish neurophysiologist: Jerzy Konorski, *Conditioned Reflexes
      and Neuron Organization* (Cambridge: Cambridge University Press, 1948).

13   Jeffrey Schwartz, an associate professor: Jeffrey M. Schwartz and Sharon Begley,
      Mind and the Brain: Neuroplasticity and the Power of Mental Force (New York:
      Regan Books, 2002), 368, 372–373.

13   Alain Brunet, an associate professor: "The Nature of Things," CBC Shows,
      November   28,   2010,   http://www.cbc.ca/video/#/Shows/The_Nature_of_
      Things/1242300217/ID=1605117929.

14   "We know the brain is like a muscle": Sophia Vinogradov, "What's New in
      Schizophrenia Research," summarized by Thomas T. Thomas, November
      28, 2007, 2. http://www.thomastthomas.com/Schizophrenia%20Research,
      Vinogradov,112807.pdf. Summary of a speech presented at the National Alli-
      ance on Mental Illness, East Bay Chapter, Albany, California, November 28,
      2007.

14   "The brain changes—physically, chemically, functionally": "The Nature of
      Things." November 28, 2010, http://www.cbc.ca/video/#/Shows/The_Nature_
      of_Things/1242300217/ID=1605117929.

CHAPTER FOUR: THE FOG

26   Of course I didn't know this at the time . . . draining the amygdala: Paul J. Whalen
      and Elizabeth A. Phelps, eds., *The Human Amygdala* (New York: Guilford Press,
      2009).

CHAPTER FIVE: BRAIN WORK: ARROWSMITH CORE PRINCIPLES

29   In 2000, Eric Kandel won the Nobel Prize: Eric Kandel, "Nobel Lecture: The
      Molecular Biology of Memory Storage: A Dialog between Genes and Synapses,"
      presented at Karolinska Institute, Stockholm, December 8, 2000. http://www
      .nobelprize.org/nobel_prizes/medicine/laureates/2000/kandel-lecture.html.

29   This was Hebb's rule: D. O. Hebb, *The Organization of Behaviour: A Neuropsycho-
      logical Theory* (New York: Wiley, 1949).

29   Mark Rosenzweig, whose work with rats so inspired me: Mark R. Rosenzweig,

"Environmental Complexity, Cerebral Change, and Behavior," *American Psychologist* 21 (1966): 321–332.

29   Subsequent animal studies: T. L. Briones, A. Y. Klintsova, and W. T. Greenough, "Stability of Synaptic Plasticity in the Adult Rat Visual Cortex Induced by Complex Environment Exposure," *Brain Research* 1018 (2004): 130–135. T. A. Comery, C. X. Stamoudis, S. A. Irwin, and W. T. Greenough, "Increased Density of Multiple-Head Dendritic Spines on Medium-Sized Spiny Neurons of the Striatum in Rats Reared in a Complex Environment," *Neurobiology of Learning and Memory* 66 (1996): 93–96. K. D. Federmeier, J. A. Kleim, and W. T. Greenough, "Learning-Induced Multiple Synapse Formation in Rat Cerebral Cortex," *Neuroscience Letters* 332 (2002): 180–184. J. A. Kleim, E. Lussnig, E. R. Schwarz, T. A. Comery, and W. T. Greenough, "Synaptogenesis and Fos Expression in the Motor Cortex of the Adult Rat after Motor Skill Learning," *Journal of Neuroscience* 16 (1996): 4529–4535.

30   But in several studies . . . an increase in gray matter: Bogdan Draganski, Christian Gaser, Gerd Kempermann, H. George Kuhn, Jürgen Winkler, Christian Büchel, and Arne May, "Temporal and Spatial Dynamics of Brain Structure Changes during Extensive Learning," *Journal of Neuroscience* 26 (2006): 6314–6317. Joenna Driemeyer, Janina Boyke, Christian Gaser, Christian Büchel, and Arne May, "Changes in Gray Matter Induced by Learning—Revisited," *PLoS One* 3 (2008): e2669. Eleanor A. Maguire, David G. Gadian, Ingrid S. Johnsrude, Catriona D. Good, John Ashburner, Richard S. J. Frackowiak, and Christopher D. Frith, "Navigation-Related Structural Change in the Hippocampi of Taxi Drivers," *Proceedings of the National Academy of Sciences, USA* 97 (2000): 4398–4403. Christian Nordqvist, "Juggling Makes Your Brain Bigger—New Study." *Medical News Today* (February 1, 2004). http://www.medicalnewstoday.com/releases/5615 .php. S. W. Lazar, C. E. Kerr, et al., "Meditation Experience Is Associated with Increased Cortical Thickness," *NeuroReport* 16 (2005): 1893–1897. Eileen Luders, Arthur W. Toga, Natasha Lepore, and Christian Gaser, "The Underlying Anatomical Correlates of Long-Term Meditation: Larger Hippocampi and Frontal Volumes of Gray Matter," *NeuroImage* 45 (2009): 672–678. J. Scholz, M. Klein, and H. Johansen-Berg, "Training-Related Cortical Thickness Changes," *Proceedings of the International Society for Magnetic Resonance in Medicine* 19 (2011): 539.

31   Rosenzweig had a sense forty-five years ago: Mark R. Rosenzweig. "Environmental Complexity, Cerebral Change, and Behavior," *American Psychologist* 21 (1966): 331.

32   Figure 2. Neuronal Plasticity: Barbro B. Johansson and Pavel V. Belichenko, "Neuronal Plasticity and Dendritic Spines: Effect of Environmental Enrichment on Intact and Postischemic Rat Brain," *Journal of Cerebral Blood Flow and Metabolism* 22 (2002): 89–96.

32   This is Luria's concept: Niels A. Lassen, David H. Ingvar, and Erik Skinhøj, "Brain Function and Blood Flow," *Scientific American* (October 1978): 70.

33   To change the brain . . . "activity-dependent neuroplasticity": Erin Clifford, "Neural Plasticity: Merzenich, Taub and Greenough," *The Harvard Brain* 6 (2009): 16–20. Mark R. Rosenzweig and Edward L. Bennett, "Psychobiology of Plasticity: Effects of Training and Experience on Brain and Behavior," *Behavioral Brain Research* 78

(1996): 57–65. Jeffrey M. Schwartz and Sharon Begley, *The Mind and the Brain: Neuroplasticity and the Power of Mental Force* (New York: Regan Books, 2002).

33 Each exercise, I also understood . . . "effortful processing": Tracey J. Shors, "Saving New Brain Cells," *Scientific American* (March 2009): 47–54. Tracey J. Shors, M. L. Anderson, D. M. Curlik, and M. S. Nokia, "Use It or Lose It: How Neurogenesis Keeps the Brain Fit for Learning," *Behavioral Brain Research* (2011) doi: 10.1016/J.bbr.2011.04.023.

34 "I do not remember exactly at what point": Jonas Salk, *Anatomy of Reality: Merging of Intuition and Reason* (New York: Praeger, 1985), 7.

36 "The degree and kind of inherent individuality": Roger Sperry, "Messages from the Laboratory," *Engineering and Science* (January 1974): 152.

CHAPTER EIGHT: LOST IN TRANSLATION

46 Luria described a problem in this part of the brain: A. R. Luria, *Traumatic Aphasia: Its Syndromes, Psychology, and Treatment*, trans. Macdonald Critchley (The Hague: Mouton & Co., 1970), 226.

46 He called it "semantic aphasia," a loss of meaning: A. R. Luria, *Basic Problems of Neurolinguistics*, trans. Basil Haigh (The Hague: Mouton & Co., 1977), 127–136, 195–201.

49 Zachary was in sensory overload: William James, *The Principles of Psychology*, vol. 1 (New York: Holt, 1890).

CHAPTER NINE: HITTING THE WALL

57 In 1978, I had been reading one of Luria's books: A. R. Luria, "Disturbances of Higher Cortical Functions with Lesions of the Frontal Region" in *Higher Cortical Functions in Man*, 2nd ed., trans. Basil Haigh (New York: Consultants Bureau, 1966, 1980), 246–365.

59 Luria defined thinking as "a special form of cognitive activity": A. R. Luria, *Human Brain and Psychological Processes*, trans. Basil Haigh (New York: Harper, 1966), 413.

62 Two Canadian neurosurgeons: Wilder Penfield and Joseph Evans, "The Frontal Lobe in Man: A Clinical Study of Maximum Removal," *Brain* 58 (1935): 115–133.

62 In 1972, psychologist Walter Mischel conducted: Jonah Lehrer, "Don't! The Secret of Self-Control," *New Yorker*, May 18, 2009.

64 According to two MIT neuroscientists: E. K. Miller and J. D. Cohen, "An Integrative Theory of Prefrontal Cortex Function," *Annual Review of Neuroscience* 24 (2001): 193.

64 A study in England in 2009: Tracy Packiam Alloway, "Working Memory, But Not IQ, Predicts Subsequent Learning in Children with Learning Difficulties," *European Journal of Psychological Assessment* 25:2 (2009): 92–98.

71 Karen came across a book on attention deficit disorder: Kate Kelly, *You Mean I'm Not Lazy, Stupid or Crazy?* (New York: Scribner, 2006).

CHAPTER TEN: WORDS FAIL

77   Luria, in *The Working Brain:* A. R. Luria, *The Working Brain* (New York: Penguin, 1973), 319.

77   In his book *Traumatic Aphasia:* A. R. Luria, *Traumatic Aphasia: Its Syndromes, Psychology, and Treatment,* trans. Macdonald Critchley (The Hague: Mouton & Co., 1970), 298–299.

86   Inner speech, Vygotsky said: Lev Vygotsky, *Thought and Language,* trans. Alex Kozulin (Cambridge, MA: MIT Press, 1986), 225, 243.

CHAPTER ELEVEN: LEAP BEFORE YOU LOOK

91   In his book: A. R. Luria, *The Working Brain* (New York: Penguin, 1973).

91   Long thought to be "lacking generally in higher cognitive function": Roger W. Sperry, "Nobel Lecture: Some Effects of Disconnecting the Cerebral Hemispheres," December 8, 1981. http://nobelprize.org/nobel_prizes/medicine/laureates/1981/sperry-lecture.html.

92   Penfield wrote: "Ruth was conscious of not being alert mentally": Wilder Penfield and Joseph Evans, "The Frontal Lobe in Man: A Clinical Study of Maximum Removals," *Brain 58* (1935): 115–133.

92   Julian Keenan, director of the Cognitive Neuroimaging Laboratory: Julian P. Keenan, Jennifer Rubio, Connie Racioppi, Amanda Johnson, and Allyson Barnacz, "The Right Hemisphere and the Dark Side of Consciousness," *Cortex* 41 (2005): 695–704.

92   Kai Vogeley, a German researcher: K. Vogeley et al., "Mind Reading: Neural Mechanisms of Theory of Mind and Self-Perspective," *NeuroImage* 14 (2001): 170–181.

101   In his book *Higher Cortical Functions in Man*: A. R. Luria, *Higher Cortical Functions in Man,* 2nd ed., trans. Basil Haigh (New York: Consultants Bureau, 1980), 335.

CHAPTER TWELVE: WHEN A PICTURE DOES NOT PAINT A THOUSAND WORDS

103   "Good-bye, till we meet again!" Lewis Carroll, *Through the Looking-Glass* (Mineola: Dover Publications, 1999), 62–63.

103   The eminent neurologist and author Oliver Sacks: Oliver Sacks, "You Look Unfamiliar," *New Yorker,* August 30, 2010, 36–43.

105   Figure 3. A model of face-processing network in the brain: David Pitcher, Vincent Walsh, and Bradley Duchaine, "The Role of the Occipital Face Area in the Cortical Face Perception Network," *Experimental Brain Research* 209 (2011): 490.

108   In her book *Visual Agnosia*: Martha J. Farah, *Visual Agnosia,* 2nd ed. (Cambridge, MA: MIT Press, 2004).

108   Research, especially Stephen Kosslyn's work: Stephen M. Kosslyn and William L. Thompson, "Shared Mechanisms in Visual Imagery and Visual Perception: Insights from Cognitive Neuroscience," in *The New Cognitive Neurosciences,* 2nd ed., ed. Michael S. Gazzaniga, 975–985 (Cambridge, MA: MIT Press, 2000).

109   What is going on, or, more correctly, what is *not* going on: Oliver Sacks, "A Man of Letters," *New Yorker,* June 28, 2010, 27.

CHAPTER THIRTEEN: A CLOSED BOOK

117   "Once you learn to read, you will be forever free": Quote attributed to Frederick Douglass (1818–1895), orator, author, reformer, and ex-slave, http://www.online-literature.com/frederick_douglass/.

118   Italian has thirty-three letter combinations: E. Paulesu et al., "Dyslexia: Cultural Diversity and Biological Unity," *Science* 291 (2001): 2165–2167. Laura Helmuth, "Neuroscience: Dyslexia: Same Brains, Different Languages," *Science* 291 (2001).

118   What exactly is dyslexia?: Sally E. Shaywitz, Maria Mody, and Bennett A. Shaywitz, "Neural Mechanisms in Dyslexia," *Current Directions in Psychological Science* 15 (2006): 278–281. Sally E. Shaywitz, "Dyslexia," *Scientific American* (November 1996): 98–104. W. E. Brown et al., "Preliminary Evidence of Widespread Morphological Variations of the Brain in Dyslexia," *Neurology* 56 (March 2001): 781–783.

119   Figure 4. The Brain and Reading: Niels A. Lassen, David H. Ingvar, and Erik Skinhøj, "Brain Function and Blood Flow," *Scientific American* (October 1978): 62–71. Julie A. Fiez and Steven E. Petersen, "Neuroimaging Studies of Word Reading," *Proceedings of the National Academy of Sciences USA* 95 (1998): 914–921. Stanislas Dehaene, *Reading in the Brain* (New York: Penguin Books, 2009).

120   In his book *Reading in the Brain*: Stanislas Dehaene, *Reading in the Brain* (New York: Penguin, 2009), 71.

122   Two American authorities on childhood development: G. R. Lyon and L. C. Moats, "Critical Conceptual and Methodological Considerations in Reading Intervention Research," *Journal of Learning Disabilities* 30 (1997): 578–588.

125   When I saw him: Steven D. Levitt and Stephen J. Dubner, *Freakonomics* (New York: HarperCollins, 2005).

CHAPTER FOURTEEN: NOTHING TO WRITE HOME ABOUT

132   Luria had a poetic phrase . . . "kinetic melody": A. R. Luria, *The Working Brain* (New York: Penguin, 1973), 32.

132   "Writing in the initial stages": Ibid.

133   In his war-wounded patients: A. R. Luria, *Higher Cortical Functions in Man,* 2nd ed., trans. Basil Haigh (New York: Consultants Bureau, 1980), 220.

134   Luria explains this: A. R. Luria, "Functional Organization of the Brain," *Scientific American* (March 1970): 69.

CHAPTER FIFTEEN: BLIND TO ONE'S OWN BODY

149   In *The Man Who Mistook His Wife for a Hat*: Oliver Sacks, "The Disembodied Lady," in *The Man Who Mistook His Wife for a Hat* (New York: Summit Books, 1985), 45–46.

153   This has been described by some clinicians . . . "afferent motor aphasia": A. R. Luria, *The Working Brain* (New York: Penguin, 1973), 174.

CHAPTER SEVENTEEN: LOST IN SPACE

168   An article in 2008 in *Science Daily* magazine: "Getting Lost: A Newly Discovered
       Developmental Brain Disorder," *Science Daily,* September 29, 2008. http://www
       .sciencedaily.com/releases/2008/09/080922135227.htm.

168   Eleanor Maguire, a professor of cognitive neuroscience: Eleanor A. Maguire et
       al., "Navigation-Related Structural Change in the Hippocampi of Taxi Drivers,"
       Proceedings of the National Academy of Sciences, USA, 97 (2000): 4398–4403.

169   Memory champions, "mental athletes" who strive to memorize: Eleanor Maguire
       et al., "Routes to Remembering: The Brains behind Superior Memory," *Nature
       Neuroscience* 6:1 (2003): 90–95.

170   Mario Lemieux once likened hockey to a game of chess: Lawrence Scanlan, *Grace
       under Fire: The State of Our Sweet and Savage Game* (Toronto: Penguin, 2002),
       128.

171   And I remember one from the 1930s: Scene from the W. C. Fields film *Man on
       the Flying Trapeze* (Paramount 1935) in *W. C. Fields: A Life on Film* by Ronald J.
       Fields (New York: St. Martin's Press, 1984). Verification courtesy of W. C. Fields
       Production, Inc.

172   Males, in fact, typically perform better than women: David C. Geary et al., "Sex
       Differences in Spatial Cognition, Computational Fluency, and Arithmetical Rea-
       soning," *Journal of Experimental Child Psychology* 77 (2000): 337–353. David C.
       Geary and Catherine DeSoto, "Sex Differences in Spatial Abilities among Adults
       from the United States and China," *Evolution and Cognition* 7 (2001): 172–177.

CHAPTER EIGHTEEN: DRAWING A BLANK

177   "Memory is intricately tied to identity": Alexandra Horowitz, "How to Memorize
       Everything," review of *Moonwalking with Einstein: The Art and Science of Remem-
       bering Everything,* by Joshua Foer, *New York Times Book Review,* March 11, 2011.
       Joshua Foer, *Moonwalking with Einstein* (New York: Penguin Press, 2011).

178   Eric Kandel, the author of *In Search of Memory:* Eric Kandel, "The Mystery of
       Memory," video, 2009. http://nobelprize.org/nobel_organizations/nobelmedia/
       partnership/s/astrazeneca/documentaries.html (accessed October 3, 2010). Eric
       R. Kandel, *In Search of Memory: The Emergence of a New Science of Mind* (New
       York: Norton, 2006).

178   "Memory is dispersed," he wrote: Anthony J. Greene, "Making Connections,"
       Scientific American Mind 24 (July–August 2009): 24.

180   In *Higher Cortical Functions in Man,* Luria describes: A. R. Luria, *Higher Cortical
       Functions in Man* (New York: Consultants Bureau, 1980), 131.

CHAPTER TWENTY: WHEN 2 + 2 DOES NOT EQUAL 4

191   "Can you do Addition?" Lewis Carroll, *Through the Looking-Glass* (Mineola:
       Dover Publications, 1999), 87–88.

192   I initially labeled a problem with number sense: A. R. Luria, *Human Brain and
       Psychological Processes,* trans. Basil Haigh (New York: Harper, 1966), 251–260.

Niels A. Lassen, David H. Ingvar, and Erik Skinhøj, "Brain Function and Blood Flow," *Scientific American* (October 1978): 70.

193   Martha Farah, founding director: Martha J. Farah, *Visual Agnosia,* 2nd ed. (Cambridge, MA: MIT Press, 2004), 8–9.

193   What Stanislas Dehaene, author of *The Number Sense:* Stanislas Dehaene et al., "Three Parietal Circuits for Number Processing," *Cognitive Neuropsychology* 20 (2003): 487–506.

194   "We cannot think about a number," says Dehaene: Stanislas Dehaene, *The Number Sense* (New York: Oxford University Press, 2011), 239.

194   When the hIPS area is not functioning properly: Brian Butterworth, "Brain's Counting Skill 'Built-In,'" *BBC News,* August 19, 2008.

CHAPTER TWENTY-ONE: IN ONE EAR AND OUT THE OTHER

201   Speech processing starts in the superior temporal region: A. R. Luria, *Higher Cortical Functions in Man,* 2nd ed., trans. Basil Haigh (New York: Consultants Bureau), 121.

201   At the age of three, a toddler's brain: Jeffrey M. Schwartz and Sharon Begley, *The Mind and the Brain: Neuroplasticity and the Power of Mental Force* (New York: Regan Books, 2002), 118–120.

202   Wepman's Auditory Discrimination Test: Joseph M. Wepman and William M. Reynolds, *Wepman's Auditory Discrimination Test,* (Los Angeles: Western Pychological Services, 1973).

203   In his book *Higher Cortical Functions in Man:* Luria, *Higher Cortical Functions in Man,* 114, 115.

CHAPTER TWENTY-TWO: THE IMPACT OF LEARNING DISABILITIES

208   Jason gave the example of an illustration: Ted Hughes, *The Iron Man: A Children's Story in Five Nights* (London: Faber and Faber, 1989).

212   How do you measure the cost when a job is lost: Burke & Associates. *A Call to Action: World Summit on LD* (Lake Louise, AB: Burke & Associates, 2008). http://www.ldac-acta.ca/en/learn-more/research/other-research.html (accessed September 15, 2010).

CHAPTER TWENTY-THREE: WORD SPREADS

215   A turning point in my life's work: Norman Doidge, "Building a Better Brain," *Saturday Night,* May 26, 2001, 22–29.

FURTHER READING

INTRODUCTION

Cole, Michael, Karl Levitin, and Alexander Luria. *The Autobiography of Alexander Luria.* Mahwah, NJ: Erlbaum, 2006.

Danforth, Scot. *The Incomplete Child: An Intellectual History of Learning Disabilities.* New York: Peter Lang, 2009.

CHAPTER ONE

Bridging the Gap between Neuroscience and Education

Atherton, Michael. "Applying the Neurosciences to Educational Research: Can Cognitive Neuroscience Bridge the Gap? Part 1." Paper presented at the annual meeting of the American Educational Research Association, Montréal, Canada, April 2005.

Bridging the Gap between Neuroscience and Education. Summary of a workshop cosponsored by the Education Commission of the States and the Charles A. Dana Foundation. Held in Denver, Colorado, July 26–28, 1996. http://www.ecs.org/clearinghouse/11/98/1198.htm.

Bruer, John T. "Education and the Brain: A Bridge Too Far." *Educational Researcher* 26:8 (1997): 4–16.

Santiago Ramón y Cajal

Llinás, Rodolfo R. "The Contribution of Santiago Ramón y Cajal to Functional Neuroscience." *Nature Reviews/Neuroscience* 4 (2003): 77–80.

Ramón y Cajal, Santiago. *Recollections of My Life.* Translated by E. Horne Craigie with Juan Cano. Cambridge, MA: MIT Press, 1989.

Neuroplasticity-Based Training with Schizophrenia

Fisher, Melissa, Christine Holland, Michael M. Merzenich, and Sophia Vinogradov. "Using Neuroplasticity-Based Auditory Training to Improve Verbal Memory in Schizophrenia." *American Journal of Psychiatry* 166:7 (2009): 805–811.

Haut, Kristen M., Kelvin O. Lim, and Angus MacDonald. "Prefrontal Cortical Changes Following Cognitive Training in Patients with Chronic Schizophrenia: Effects of Practice, Generalization, and Specificity." *Neuropsychopharmacology* 35:9 (2010): 1850–1859.

CHAPTER FIVE

Neuroplasticity

Auld, Daniel S., and Richard Robitaille. "Glial Cells and Neurotransmission: An Inclusive View of Synaptic Function." *Neuron* 40 (2003): 389–400.

Finlay, Barbara L., Richard B. Darlington, and Nicholas Nicastro. "Developmental Structure in Brain Evolution." *Behavioral and Brain Sciences* 24 (2001): 263–308.

Klintsova, A. Y., and W. T. Greenough. "Synaptic Plasticity in Cortical Systems." *Current Opinion in Neurobiology* 9 (1991): 203–208.

Marty, Serge, Maria da Beringer, and Benedikt Berninger. "Neurotrophins and Activity-Dependent Plasticity of Cortical Interneurons." *Trends in Neuroscience* 20 (1997): 198–202.

CHAPTER EIGHT

ADD and ADHD: Link to Prefrontal Cortex

Fuster, Joaquin M. *The Prefrontal Cortex,* 4th ed. London: Academic Press, 2009.

Makris, Nikos, Stephen L. Buka, Joseph Biederman, George M. Papadimitriou, Steven M. Hodge, Eve M. Valera, et al. "Attention and Executive Systems Abnormalities in Adults with Childhood ADHD: A DT-MRI Study of Connections." *Oxford Journals* 18 (2007): 1210–1220.

Monuteaux, Michael C., Larry J. Siedman, Stephen V. Faraone, Nikos Makris, Thomas Spencer, Eve Valera, Ariel Brown, et al. "A Preliminary Study of Dopamine D4 Receptor Genotype and Structural Brain Alterations in Adults with ADHD." *American Journal of Medical Genetics Part B* (Neuropsychiatric Genetics) 147B (2008): 1436–1441.

Raz, Amir. "Brain Imaging Data of ADHD." *Psychiatric Times* 11:9 (2004): 1–4.

Valera, Eve M., Stephen V. Faraone, Kate E. Murray, and Larry J. Seidman. "Meta-Analysis of Structural Imaging Findings in Attention-Deficit/Hyperactivity Disorder." *Society of Biological Psychiatry* 61 (2007): 1361–1369.

ADD and ADHD: Link to Subcortical Regions and Neurochemical Factors

Aguiar, Andrea, Paul A. Eubig, and Susan L. Schantz. "Attention Deficit/Hyperactivity Disorder: A Focused Overview for Children's Environmental Health Researchers." *Environmental Health Perspectives* 118:12 (2010): 1–51.

Amat, Jose A., Richard A. Bronen, Sanjay Salua, Noriko Sato, Hongtu Zhu, Daniel Gorman, Jason Royal, and Bradley S. Peterson. "Increased Number of Subcortical Hyperintensities on MRI in Children and Adolescents with Tourette's Syndrome, Obsessive-Compulsive Disorder, and Attention Deficit Hyperactivity Disorder." *American Journal of Psychiatry* 163 (2006): 1106–1108.

Ernst, Monique, and Alan Zametkin. "The Interface of Genetics, Neuroimaging, and Neurochemistry in Attention-Deficit Hyperactivity Disorder." In *Psychopharmacol-*

ogy: The Fourth Generation of Progress. Edited by F. Bloom and D. Kupfer. New York: Raven Press, 1995, 1643–1652.

Geller, Barbara. "Neurochemistry of ADHD and Its Medications." *Journal Watch Psychiatry,* April 9, 2003.

Lopez-Larson, M., E. S. Michael, J. E. Terry, J. L. Breeze, S. M. Hodge, L. Tang, D. N. Kennedy, et al. "Subcortical Differences among Youths with Attention-Deficit/ Hyperactivity Disorder Compared to Those with Bipolar Disorder with and without Attention-Deficit/Hyperactivity Disorder." *Journal of Child and Adolescent Psychopharmacology* 19:1 (2009): 31–39.

McKay, K. E., and J. M. Halperin. "ADHD, Aggression, and Antisocial Behavior across the Lifespan: Interactions with Neurochemical and Cognitive Function." *Annals of the New York Academy of Sciences* 931 (2001): 84–96.

Voeller, Kytja K. S. "Attention-Deficit Hyperactivity Disorder (ADHD)." *Journal of Child Neurology* 19 (2004): 798–814.

Volkow, Nora D., Gene-Jack Wang, Joanna A. Fowler, Jean Logan, Madina Gerasimov, Laurence Maynard, Yu-Shin Ding, et al. "Therapeutic Doses of Oral Methylphenidate Significantly Increase Extracellular Dopamine in the Human Brain." *Journal of Neuroscience* 21 (2001): 1–5.

CHAPTER NINE

Stanford Marshmallow Experiment

Lehrer, Jonah. "The Willpower Circuit." www.wired.com/wiredscience/frontal-cortex (accessed September 6, 2011).

CHAPTER ELEVEN

The Prefrontal Cortex in Social Cognition

Amodio, David M., and Chris D. Frith. "Meeting of Minds: The Medial Frontal Cortex and Social Cognition." *Nature Reviews Neuroscience* 7 (2006): 268–277.

Feinberg, Todd E., and Julian Paul Keenan. "Where in the Brain Is the Self?" *Conscious Cognition* 14 (2005): 661–678.

Forbes, Chad E., and Jordan Grafman. "The Role of the Human Prefrontal Cortex in Social Cognition and Moral Judgment." *Neuroscience* 33 (2010): 299–324.

Frith, Chris, and Daniel Wolpert. *The Neuroscience of Social Interaction: Decoding, Imitating and Influencing the Actions of Others.* New York: Oxford University Press 2003.

Goel, Vinod, and Jordan Grafman. "Role of the Right Prefrontal Cortex in Ill-Structured Planning." *Cognitive Neuropsychology* 17 (2000): 415–436.

Jimura, Koji, Seiki Konishi, and Yasushi Miyashita. "Dissociable Concurrent Activity of Lateral and Medial Frontal Lobe during Negative Feedback Processing." *NeuroImage* 22 (2004): 1578–1586.

Keenan, Julian Paul, Mark A. Wheeler, Gordon G. Gallup, Jr., and Alvaro Pascual-Leone.

"Self-Recognition and the Right Prefrontal Cortex." *Trends in Cognitive Sciences* 4 (2000): 338–344.

Knoch, Daria, Alvaro Pascual-Leone, Kasper Meyer, Valerie Treyer, and Ernst Fehr. "Diminishing Reciprocal Fairness by Disrupting the Right Prefrontal Cortex." *Science* 314 (2006): 829–832.

Mason, Robert, and Marcel Just. "The Role of the Theory-of-Mind Cortical Network in the Comprehension of Narratives." *Language and Linguistics Compass* 3:1 (2009): 157–174.

Singer, Tania. "Understanding Others: Brain Mechanisms of Theory of Mind and Empathy." In *Neuroeconomics: Decision Making and the Brain.* Edited by P. W. Glimcher, C. F. Camerer, E. Fehr, and R. A. Poldrack. Amsterdam: Elsevier, 2008, pp. 251–268.

Seitz, Rudiger J., Janpeter Nickel, and Nina P. Azari. "Functional Modularity of the Medial Prefrontal Cortex: Involvement in Human Empathy." *Neuropsychology* 20 (2006): 743–751.

Shallice, Tim. "'Theory of Mind' and the Prefrontal Cortex." *Brain* 124 (2001): 247–248.

Stuss, Donald T., Gordon G. Gallup, and Michael P. Alexander. "The Frontal Lobes Are Necessary for 'Theory of Mind.'" *Brain* 124 (2001): 279–286.

Vollm, Birgit A., Alexander N. W. Taylor, Paul Richardson, Rhiannon Corcoran, John Stirling, Shane McKie, John F. W. Deakin, and Rebecca Elliott. "Neuronal Correlates of Theory of Mind and Empathy: A Functional Magnetic Resonance Imaging Study in a Nonverbal Task." *NeuroImage* 29 (2006): 90–98.

CHAPTER TWELVE

Visual Imagery and Perception

Borst, Gregoire, and Stephen M. Kosslyn. "Visual Mental Imagery and Visual Perception: Structural Equivalence Revealed by Scanning Processes." *Memory and Cognition* 36 (2009): 849–862.

Kosslyn, Stephen M. "Remembering Images." In *Memory and Mind.* Edited by Mark A. Gluck, John R. Anderson, and Stephen M. Kosslyn. New York: Taylor and Francis, 2008, pp. 93–109.

Slotnick, Scott D., William L. Thompson, and Stephen M. Kosslyn. "Visual Mental Imagery Induces Retinotopically Organized Activation of Early Visual Areas." *Cerebral Cortex* 15 (2005): 1570–1583.

Face Perception and Object Recognition

Banissy, Michael, Lúcia Garrido, Flor Kusnir, Bradley Duchaine, Vincent Walsh, and Jamie Ward. "Superior Facial Expression, But Not Identity Recognition, in Mirror-Touch Synesthesia." *Journal of Neuroscience* 51 (2011): 1820–1824.

Calder, Andrew J., Gillian Rhodes, Mark H. Johnson, and James V. Haxby, eds. *The Oxford Handbook of Face Perception.* New York: Oxford University Press, 2011.

Calder, Andrew J., and Andrew W. Young. "Understanding the Recognition of Facial Identity and Facial Expression." *Nature Reviews Neuroscience* 6 (2005): 641–651.

Engell, Andrew D., and James V. Haxby. "Facial Expression and Gaze-Direction in Human Superior Temporal Sulcus." *Neuropsychologia* 45 (2007): 3234–3241.

Evans, J. J., A. J. Heggs, N. Antoun, and J. R. Hodges. "Progressive Prosopagnosia Associated with Selective Right Temporal Lobe Atrophy." *Brain* 118 (1) (1995): 1–13.

Grill-Spector, Kalanit, Nicholas Knouf, and Nancy Kanwisher. "The Fusiform Face Area Subserves Face Perception Not Generic within-Category Identification." *Nature Neuroscience* 7 (2004): 555–562.

Grill-Spector, Kalanit, and Rory Seyres. "Object Recognition: Insights from Advances in fMRI Methods." *Current Directions in Psychological Science* 17:2 (2008): 73–79.

Haxby, James V., Elizabeth A. Hoffman, and M. Ida Gobbini. "The Distributed Human Neural System for Face Perception." *Trends in Cognitive Science* 4 (6) (2000): 223–233.

Hussey, Elizabeth, and Ashley Safford. "Perception of Facial Expression in Somatosensory Cortex Supports Simulationist Models." *Journal of Neuroscience* 29:2 (2009): 301–302.

Pancarogl, Raika, Thomas Busigny, Samantha Johnston, Alla Sekunova, Bradley Duchaine, and Jason J. S. Barton. "The Right Anterior Temporal Lobe Variant of Prosopagnosia." *Journal of Vision* 11:11 (2011): Article 573.

Pitcher, David, Lucie Charles, Joseph T. Devlin, Vincent Walsh, and Bradley Duchaine. "Triple Dissociation of Faces, Bodies, and Objects in Extrastriate Cortex." *Current Biology* 19:4 (2009): 319–324.

Pitcher, David, Daniel D. Dilks, Rebecca R. Saxe, Christina Triantafyllou, and Nancy Kanwisher. "Differential Selectivity for Dynamic Versus Static Information in Face-Selective Cortical Regions." *NeuroImage* 56 (2011): 2356–2363.

Pitcher, David, Lúcia Garrido, Vincent Walsh, and Bradley C. Duchaine. "Transcranial Magnetic Stimulation Disrupts the Perception and Embodiment of Facial Expressions." *Journal of Neuroscience* 28 (2008): 8929–8933.

Said, Christopher P., Christopher D. Moore, Kenneth D. Norman, James V. Haxby, and Alexander Todorov. "Graded Representations of Emotional Expressions in the Left Superior Temporal Sulcus." *Frontiers in Systems Neuroscience* 4:6 (2010): 1–8.

The Visual Word Form Area, the Brain's Letterbox, and Symbol Recognition

Cohen, Laurent, Stanislas Dehaene, Lionel Naccache, Stephane Lehericy, Ghislaine Dehaene-Lambertz, Marie-Anne Henaff, and François Michel. "The Visual Word Form Area: Spatial and Temporal Characterization of an Initial Stage of Reading in Normal Subjects and Posterior Split-Brain Patients." *Brain* 123 (2000): 291–307.

Cohen, Laurent, and Stanislas Dehaene. "Specialization within the Ventral Stream: The Case for the Visual Word Form Area." *NeuroImage* 22 (2004): 466–476.

Dehaene, Stanislas, Gurvan Le Clec'H, Jean-Baptiste Poline, Denis Le Bihan, and Laurent Cohen. "The Visual Word Form Area: A Prelexical Representation of Visual Words in the Fusiform Gyrus." *NeuroReport* 13 (2002): 321–325.

Polk, Thad A., and Martha J. Farah. "The Neural Development and Organization of Letter Recognition: Evidence from Functional Neuroimaging, Computational Modeling, and Behavioral Studies." *Proceedings of the National Academy of the Sciences USA* 95: 3 (1998): 847–852.

Broca's Area

Burton, Martha W., Steven L. Small, and Sheila E. Blumstein. "The Role of Segmentation in Phonological Processing: An fMRI Investigation." *Journal of Cognitive Neuroscience* 12 (2000): 679–690.

Cornelissen, Piers L., Morten L. Kringelbach, Andrew W. Ellis, Carol Whitney, Ian E. Holliday, and Peter C. Hansen. "Activation of the Left Inferior Frontal Gyrus in the First 200 ms of Reading: Evidence from Magnetoencephalography (MEG)." *PLoS One* 4: 4 (2009): 1–25.

Gernsbacher, Morton Ann, and Michael P. Kaschak. "Neuroimaging Studies of Language Production and Comprehension." *Annual Review of Psychology* 54 (2002): 94–114.

Hagoort, Peter, and Willem J. M. Levelt. "The Speaking Brain." *Science* 326 (2009): 372–373.

Heim, Stefan, Kai Alter, Anja K. Ischebeck, Katrin Amunts, Simon B. Eickhoff, Harmut Mohlberg, Karl Zilles, D. Yeves von Cramon, and Angela D. Friederic. "The Role of the Left Brodman's Areas 44 and 45 in Reading Words and Pseudowords." *Cognitive Brain Research* 25 (2005): 982–993.

CHAPTER FOURTEEN

Neural Substrates of Writing

Beeson, Pelagie M., Steven Z. Rapcsak, Elena Plante, Jullyn Chargualaf, Anne Chung, Sterling C. Johnson, and Theodore P. Trouard. "The Neural Substrates of Writing: A Functional Magnetic Resonance Imaging Study." *Aphasiology* 17 (2003): 647–665.

MacArthur, Charles A., Steve Graham, and Jill Fitzgerald. *Handbook of Writing Research*. New York: Guilford Press, 2006.

Pierrot-Deseilligny, Charles, Dan Milea, and Rene M. Muri. "Eye Movement Control by the Cerebral Cortex." *Current Opinion in Neurology* 17 (2004): 17–25.

CHAPTER FIFTEEN

Role of Somatosensory Cortex in Kinesthetic Perception

Azañón, Elena, Matthew R. Longo, Salvador Soto-Faraco, and Patrick Haggard. "The Posterior Parietal Cortex Remaps Touch into External Space." *Current Biology* 20 (2010): 1304–1309.

Bernier, Pierre-Michel, and Scott T. Grafton. "Human Posterior Parietal Cortex Flexibly

Determines Reference Frames for Reaching Based on Sensory Context." *Neuron* 68 (2010): 776–788.

Binkofski, F., E. Kunesch, J. Classen, R. J. Seitz, and H. J. Freund. "Tactile Apraxia: Unimodal Apractic Disorder of Tactile Object Exploration Associated with Parietal Lobe Lesions." *Brain* 124 (2001): 132–144.

Feldman, Anatol G. "Threshold Position Control Signifies a Common Special Frame of Reference for Motor Action and Kinesthesia." *Brain Research Bulletin* 75 (2008): 497–499.

Homke, Lars, Katrin Amunts, Lutz Bönig, Christian Fretz, Ferdinand Binkofski, Karl Zilles, and Bruno Weder. "Analysis of Lesions in Patients with Unilateral Tactile Agnosia Using Cytoarchitectonic Probabilistic Maps." *Human Brain Mapping* 30 (2009): 1444–1456.

Larner, Andrew J. "Braille Alexia: An Apperceptive Tactile Agnosia?" *Journal of Neurology, Neurosurgery and Psychiatry* 78 (2007): 907–908.

Naito, Eiichi, Per E. Roland, Christian Grefkes, H. J. Choi, Simon Eickoff, Stefan Geyer, Karl Zilles, and H. Henrik Ehrsson. "Dominance of the Right Hemisphere and Role of Area 2 in Human Kinesthesia." *Journal of Neurophysiology* 93 (2005): 1020–1034.

Reed, Catherine L., Shy Shoham, and Eric Halgren. "Neural Substrates of Tactile Object Recognition: An fMRI Study." *Human Brain Mapping* 21 (2004): 236–246.

Speech Perception and Production and the Somatosensory Cortex

Houde, John F. "There's More to Speech Perception Than Meets the Ear." *Proceedings of the National Academy of Sciences, USA* 106 (2009): 20139–20140.

Ito, Takayuki, Mark Tiede, and David J. Ostry. "Somatosensory Function in Speech Perception." *Proceedings of the National Academy of Sciences, USA* 106 (2009): 1245–1248.

Nasir, Sazzad M., and David J. Ostry. "Somatosensory Precision in Speech Production." *Current Biology* 16 (2006): 1918–1923.

Trembley, Stephanie, Douglas M. Shiller, and David J. Ostry. "Somatosensory Basis of Speech Production." *Nature* 423 (2003): 866–869.

CHAPTER SEVENTEEN

Spatial Cognition

Arleo, A., and L. Rondi-Reig. "Multimodal Sensory Integration and Concurrent Navigation Strategies for Spatial Cognition in Real and Artificial Organisms." *Journal of Integrative Neuroscience* 6 (2007): 327–366.

Bilalic, Merim, Andrea Kiesel, Carsten Pohl, Michael Erb, and Wolfgang Grodd. "It Takes Two: Skilled Recognition of Objects Engages Lateral Areas in Both Hemispheres." *PLoS One* 6 (2011): e16202.

Broadbent, Nicola J., Larry R. Squire, and Robert E. Clark. "Spatial Memory, Recognition

Memory, and the Hippocampus." *Proceedings of the National Academy of Sciences USA* 101 (2004): 14515–14520.

Calton, Jeffrey L., and Jeffrey S. Taube. "Where Am I and How Will I Get There from Here? A Role for Posterior Parietal Cortex in the Integration of Spatial Information and Route Planning." *Neurobiology of Learning and Memory* 91 (2009): 186–196.

Holscher, Christoph, Thomas F. Shipley, Marta Olivetti Belardinelli, John A. Batement, and Nora S. Sewcombe, eds. *Spatial Cognition VII.* New York: Springer, 2010.

Iaria, Giuseppe, and Jason J. S. Barton. "Developmental Topographical Disorientation: A Newly Discovered Cognitive Disorder." *Experimental Brain Research* 206 (2010): 189–196.

Iaria, Giuseppe, Nicholas Bogod, Christopher J. Fox, and Jason J. S. Barton. "Developmental Topographical Disorientation: Case One." *Neuropsychologia,* 47:1 (2009), 30–40.

Iaria, Giuseppe, Liana Palermo, Giorgia Committeri, and Jason J. S. Barton. "Age Differences in the Formation and Use of Cognitive Maps." *Behavioural Brain Research* 196 (2008): 187–191.

Kumaran, Dharshan, and Eleanor A. Maguire. "The Human Hippocampus: Cognitive Maps or Relational Memory?" *Journal of Neuroscience* 25 (2005): 7254–7259.

Lim, Tae-Sung, Giuseppe Iaria, and So Young Moon. "Topographical Disorientation in Mild Cognitive Impairment: A Voxel-Based Morphometry Study." *Journal of Clinical Neurology* 6 (2010): 204–211.

Maguire, Eleanor A., Neil Burgess, James G. Donnett, Richard S. J. Frackowiak, Christopher D. Frith, and John O'Keefe. "Knowing Where and Getting There: A Human Navigation Network." *Science* 280 (5365) (1998): 921–924.

Mast, Fred W., and Lutz Jancke, eds. *Spatial Processing in Navigation, Imagery and Perception.* New York: Springer, 2007.

Nico, D., L. Piccardi, G. Iaria, F. Bianchini, L. Zompanti, and C. Guariglia. "Landmark Based Navigation in Brain-Damaged Patients with Neglect." *Neuropsychologia* 46 (2008): 1898–1907.

Piccardi, Laura, Giuseppe Iaria, Maura Ricci, Filippo Bianchini, Laura Zompanti, and Cecilia Guariglia. "Walking in the Corsi Test: Which Type of Memory Do You Need?" *Neuroscience Letters* 432 (2008): 127–131.

van der Ham, Ineke J. M., Martine J. E. van Zandvoort, Tobias Meilinger, Sander E. Bosch, Neeltje Kant, and Albert Postma. "Spatial and Temporal Aspects of Navigation in Two Neurological Patients." *NeuroReport* 21 (2010): 685–689.

Weiss, E., C. M. Siedentopf, A. Hofer, E. A. Deisenhammer, M. J. Hoptman, C. Kremser, S. Golaszewski, S. Felber, W. W. Fleischhacker, and M. Delazer. "Sex Differences in Brain Activation Pattern during a Visuospatial Cognitive Task: A Functional Magnetic Resonance Imaging Study in Healthy Volunteers." *Neuroscience Letters* 334 (2003): 169–172.

Wright, Rebecca, William L. Thompson, Giorgio Ganis, Nora S. Newcombe, and Stephen M. Kosslyn. "Training Generalized Spatial Skills." *Psychonomic Bulletin and Review* 15 (2008): 763–771.

CHAPTER EIGHTEEN

Schacter, Daniel L. *The Seven Sins of Memory.* New York: Houghton Mifflin, 2011.

CHAPTER TWENTY

Mathematical Cognition

Butterworth, Brian. *The Mathematical Brain.* London: Macmillan, 1999.

Campbell, Jamie I. D., ed. *Handbook of Mathematical Cognition.* New York: Psychology Press, 2005.

Delazer, M., F. Domahs, L. Bartha, C. Brenneis, A. Lochy, T. Trieb, and T. Benke. "Learning Complex Arithmetic—an fMRI Study." *Cognitive Brain Research* 18 (1) (2003): 76–88.

Grabner, Roland H., Daniel Ansari, Gernot Reishofer, Elizabeth Stern, Franz Ebner, and Christa Neuper. "Individual Differences in Mathematical Competence Predict Parietal Brain Activation during Mental Calculation." *NeuroImage* 38 (2007): 346–356.

Ischebeck, Anja, Laura Zamarian, Michael Schocke, and Margarete Delazer. "Flexible Transfer of Knowledge in Mental Arithmetic: An fMRI Study." *NeuroImage* 44 (2009): 1103–1112.

Landgraf, S., E. van der Meer, and F. Krueger. "Cognitive Resource Allocation for Neural Activity Underlying Mathematical Cognition: A Multi-Method Study." *International Journal of Mathematics Education* 42 (2010): 579–590.

Molko, N., A. Cachia, D. Riviere, J. F. Bruandet, D. Le Bihan, D. Cohen, and S. Dehaene. "Functional and Structural Alterations of the Intraparietal Sulcus in a Developmental Dyscalculia of Genetic Origin." *Neuron* 40 (2003): 847–858.

CHAPTER TWENTY-ONE

Studies Demonstrating Speech-Sound-Sensitive Areas in the Superior Temporal Region (Superior Temporal Gyrus and Superior Temporal Sulcus)

Belin, P., R. J. Zatorre, P. Lafaille, P. Ahad, and B. Pike. "Voice-Selective Areas in Human Auditory Cortex." *Nature* 403 (2000): 309–312.

Binder, J. R., J. A. Frost, T. A. Hammeke, P. S. Bellgowan, J. A. Springer, J. N. Kaufman, and E. T. Possing. "Human Temporal Lobe Activation by Speech and Nonspeech Sounds." *Cerebral Cortex* 10 (2000): 512–528.

Boatman, Dana F., and Diana L. Miglioretti. "Cortical Sites Critical for Speech Discrimination in Normal and Impaired Listeners." *Journal of Neuroscience* 25 (2005): 5475–5480.

Buchsbaum, Bradley R., Gregory Hickok, and Colum Humphries. "Role of Left Posterior Superior Temporal Gyrus in Phonological Processing for Speech Perception and Production." *Cognitive Science* 25 (2001): 663–668.

Hickok, G., and D. Poeppel. "The Cortical Organization of Speech Processing." *Nature Review Neuroscience* 8 (2007): 393–402.

Kovelman, Ioulia, Jonathan C. Yip, and Erica L. Beck. "Cortical Systems That Process Language, as Revealed by Non-Native Speech Sound Perception." *Neuroreport* 22 (2011): 947–950.

Liebenthal, E., J. R. Binder, S. M. Spitzer, E. T. Possing, and D. A. Medler. "Neural Substrates of Phonemic Perception." *Cerebral Cortex Journal* 15 (2005): 1621–1631.

Scott, S. K., and I. S. Johnsrude. "The Neuroanatomical and Functional Organization of Speech Perception." *Trends in Neuroscience* 26 (2003): 100–107.

ACKNOWLEDGMENTS

This book owes a debt of gratitude to many people:

My father and mother, who created the nature and nurture from which this work grew.

My brothers, Alex, Greg, Donald, and Will, and my sisters-in-law, Michelle, Anne, Jennifer, and Therese, who have supported me with love and encouragement.

And a special thanks to my brother Donald, who has been part of this work with me since the early 1980s and translated many of the cognitive exercises into computer programs.

The Arrowsmith community—all the voices of all the individuals who have come through the doors of Arrowsmith, some of whom shared their stories for this book, for all that you have taught and continue to teach me, and to all the teachers and administrators who have implemented this program in their schools.

All of the Arrowsmith staff, who every day make this work manifest: Andrea Peirson, Christina Furtado, Helen Fadakis, Julia Reynolds-McNeill, Lise Waller, Donald Young, William Young, Janice Jordan, Kirsti Jussila, Ian Taylor-Wright, Sherri Lane Howie, Michael Cerovich, Jason Kinsey, Stacey Hobbs, Sarah Maltby, Daina Luszczek, Audrey Howard, Anna Calleja, Adriene Pratt, Jennifer Richmond, Michelle Roach, Jennifer Smith, Shannon Duke, Ivana Velimirovic, Matthew Coppins, Rob Gunning, Christine Hopkins, Kayla Gunning, and Jennifer Scates.

Tara Anchel, assistant director of Arrowsmith School in Toronto, and Jill Marcinkowski, assistant director of Arrowsmith School Peterborough, for so capably and competently running the two schools and for the tremendous heart and compassion you each embody and that is so present in your work.

Annette Goodman, chief education officer of Arrowsmith Program, you bring your intelligence and your heart to this work, and you listen deeply to the concerns of educators, parents, and students and utilize this information to enhance the work. And my thanks for the deep personal connection that has continued and grown since we first met in 1991.

The teachers who first implemented this program beyond the walls of Arrowsmith School starting in 1997, and several who still do: Sheila Brown, Marg Quinn, Mary Jane McKeen, Renata DiPiero, Carmela Ferri, Elaine Magennis-Hill, Sharron Rose, Erica Powers, and Mary Feliciani. And two administrators who supported this early venture: Johanne Stewart (former Director of Education of the Toronto Catholic District School Board) and Sandra Montgomery (former Superintendent of Special Services).

Another first, Rabbi Dr. Heshy Glass, the first principal in the United States to implement the Arrowsmith Program.

Norman Doidge for seeing many years ago the possibility of healing in the work of Arrowsmith for people whose learning disabilities lead them to the psychiatrist's couch, and for educating people on a range of practical applications of neuroplasticity that can lead to healing of conditions that were once believed to be fixed and intractable.

Howard Eaton for being intellectually curious upon seeing significant cognitive change in several of his clients as the result of this work, leading to his researching this new paradigm and then implementing it and writing a book, *Brain School*, on the results.

William Lancee, who has given advice on research design and methods of data analysis for the studies conducted on the outcome of the program and who has worked on converting our paper-and-pencil assessment into a web-based assessment.

Jackie Kaiser, my agent at Westwood Creative Artists, who provided support, guidance, and wise counsel, and said this book needed to be written.

Hilary Redmon for her judicious editing that illuminated what was essential in the stories.

Sydney Tanigawa, who helped nurture this book through its final critical stages while embodying grace under pressure.

Marta Scythes for her wonderful illustrations.

Ryk Narrel for transforming my old photographs.

Larry Scanlan for the calm and levelheadedness you brought to the writing process.

Ken Rose and Michael Moskowitz for their suggestions and corrections to the description of neural transmission in Chapter 5.

Barbro Johansson for generously sharing her slides demonstrating what happens to a neuron as a result of stimulation, which became the basis for the illustration in Chapter 5.

David Pitcher for kindly allowing me to adapt a slide from his research outlining several of the cortical areas involved in face perception.

Nechama Karman for generously answering my questions on sensory feedback.

Michael Cole for verifying facts about Luria.

Catherine Roe for helping with the important behind-the-scenes details, including finding and documenting all the references.

Freyda Isaacs for your fierce compassion, for many years of guidance, and for supporting me on the roller-coaster ride as this book was birthed.

Marla Golden and Michael Moskowitz for your brilliant work applying the principles of neuroplasticity in service of healing chronic pain. You have given me back my life and the energy to complete this book.

Ellen Cutler for your work with enzymes, which allowed me to get sustenance.

Barbara Burke for being my friend through the past fifteen years, as well as my spiritual adviser who listens to and supports all aspects of me.

My friends who encouraged me as I wrote this book: Carla Peppler, Marsha Stonehouse, Julaine Brent, Helen Valleau, Lynda Visosky, Tammy Felts, Wendy Abbott, Steffany Hanlen, Janice Mahwinney, Shari Ezyk, John Ezyk, Iryna Bonya, Andrew Putintsev, Devorah Garland, Vikki Foy, and Don Frost.

A special thank you to Barbara Burke, Kimberly Carroll, Aline Couture, Dorothy Holden, Rebecca Parkes, Neil Schwartzbach, Sally Schwartzbach, and Bruce Schwartzentruber for reading an early version of the manuscript and providing feedback.

And last—without whom this work would never have been developed and this book could never have been written—all the neuroscientists who have made it their life's work to aid our understanding of the brain. A special thanks to A. R. Luria for shining a light on my darkness and Mark Rosenzweig for showing me the possibility of a way out of the darkness.

I apologize for those whom I have missed. All errors and omissions are mine.

INDEX

Page numbers in italics refer to figures.

Abstract reasoning deficit, 222–23
Accident proneness, 5, 101. *See also*
    Clumsiness
Activity-dependent neuroplasticity, 14, 33
ADHD. *See* Attention deficit hyperactivity
    disorder
Afferent motor aphasia, 153
Alzheimer's patients, 168
American Christian School, 138, 207
Amygdala, 26, 189
*Anatomy of Reality: Merging of Intuition
    and Reason* (Salk), 34
Anchel, Tara, 209
Andreou, Dimitri, 97
Andreou, Elena, 97
Angular gyrus, 47, 194, 228
Animal studies, 7, 13, 29–30, 31
Anterior temporal lobe, *105*
Apgar scores, 84
Aphasia
    afferent motor, 153
    semantic, 46
Apraxia, positional, 153
Arlington, Amber, 183–84, 192, 193
Arlington, Mary, 183–84, 191–92, 193
Arrowsmith, Louie May (grandmother), 16
Arrowsmith Auditory Speech
    Discrimination Test, 202
Arrowsmith Program
    for artifactual thinking deficit, 102
    for auditory speech discrimination
        deficit, 204
    core principles of, 29–36
    description of cognitive deficits
        addressed by, 217–23
    fundamental premise of, 11
    for kinesthetic perception deficit,
        153–55
    for memory-for-information deficit,
        180, 182
    for motor symbol sequencing deficit,
        135–37
    number of programs, 216
    for object recognition deficit, 112
    for predicative speech deficit, 81,
        84–85, 87
    for primary motor deficit, 159
    for quantification deficit, 198–99
    for reading dysfunction, 120, 123, 128
    for spatial reasoning deficit, 174
    for symbolic thinking deficit, 58–59,
        60–61, 71, 102
    for symbol relations deficit, 50, 51,
        52, 55
Arrowsmith School, 36, 124, 207, 209
    Chinese ideogram for courage in, 211
    Doidge on, xiv–xvii, 215
    expansion of, 215–16
    folding of, 187, 190
    launching of, 11, 161–63
    New York version of, 162–63, 187,
        190
    relaunching of, 190
Arrowsmith-Young, Barbara
    birth of, 15
    childhood of, 16–17
    congenital asymmetry of, 15
    depression suffered by, 6, 26, 207–8
    Doidge on accomplishments of,
        xiv–xvii
    early education of, 4–6, 19–21
    emotional upheaval in life of, 187–90
    family background of, 15–16
    fog experienced by, 4, 23–27
    graduate school education of, 38–39
    high school education of, 23–27, 37

Arrowsmith-Young, Barbara (*cont.*)
  identification with Zazetsky, 4, 5, 6–7,
      26
  illnesses of, 39, 190
  kinesthetic perception deficit of, 5, 6,
      16, 145–48, 172
  kinesthetic speech disorder of, 152
  marriage of, 39–40
  master's thesis of, 161–62, 167
  neurological deficits shared with
      parents, 42–43
  personal use of cognitive exercises, 7–8,
      41–44, 147–48, 162, 188
  progress made by, 41–44, 148, 175–76
  spatial reasoning deficit of, 146–47,
      165–68, 170–71, 175–76
  symbol relations deficit of, 24, 45, 147,
      188
  undergraduate education of, 37–38
Artifactual thinking deficit, 89–102, 217
attentional difficulties and, 55, 100
  Brodmann area corresponding to, 228
  cognitive exercises for, 102
  description of, 221
  emotional intelligence and, 93, 102
  executive role of, 92, 96
  to experience, 90
  features of, 89, 92–93, 96, 98–99
  hallmark feature of, 90–91
  intelligence/IQ and, 102
  no rule book for life and, 102
  self-awareness and, 92–93
  social interaction and, 90–98
  symbolic thinking deficit and, 91,
      96–97, 102
  theory of mind and, 92–93
Assembly, difficulties with, 171–72, 175
Athleticism. *See* Sports
Attentional problems, 54–55, 100–102
Attention deficit disorder, 71, 96
Attention deficit hyperactivity disorder
      (ADHD), 54–55, 63, 100
Auditory processing problem, 49
Auditory speech discrimination deficit,
      120, 201–5
  Brodmann area corresponding to, 228
  cognitive exercises for, 204

description of, 219
  distinguishing phonemic signs, 203
  to experience, 202
  features of, 203–4
Axons, 30, *31*
Axon terminals, 30, *31*

Badminton, 147, 153, 154
BDNF (brain-derived neurotropic factor),
      14
Body blindness. *See* Kinesthetic perception
      deficit
Body language, inability to read. *See*
      Artifactual thinking deficit
"Book of Hours, The" (Levertov), x
Brain. *See also* Neuroplasticity; specific
      regions of
  Brodmann areas of, 227–28
  complexity of, 36
  exercise impact on, 13
  face-processing network in, *105*
  hardwired-machine model of, xiii, 5,
      9, 10
  lobes of, *225*
  reading and, *119*
  restructuring of auditory networks, 201
Brain-derived neurotropic factor (BDNF),
      14
Brain imaging. *See* Neuroimaging studies
Brain School (Eaton), xvii
Brain's letterbox, 121. *See also* Occipital-
      temporal area of the brain
Brain That Changes Itself, The (Doidge),
      xiv, 10
Broca, Pierre Paul, 121, 219
Broca's area of the brain, *119,* 120, 123,
      127, 130, 219
  author's weakness in, 42–43
  cognitive exercise for, 128
  to experience, 121–22
  language-central role of, 121–22
Broca's speech pronunciation, 219
Brodmann, Korbinian, 227
Brodmann areas of the brain, *119, 227,*
      227–28
Brunet, Alain, 13
"Building a Better Brain" (Doidge), 215

Bullying, 94, 208
Bundy, Rex, 60, 162
Bush, George H. W., 12
Butterworth, Brian, 194

Cab drivers, brains of. *See* London cab
    drivers, brains of
Carroll, Lewis, 103, 191
Cell body, *31*
Center for Cognitive Neuroscience, 193
Cerebellum, 13
Cerebral palsy, 126
Chess, 170
Cicada metaphor, 207
Cicero, 152
Clock-reading exercise, 138
    author's development and use of, 7–8,
        41, 42, 162
    for symbol relations deficit, 50, 51,
        52, 55
Clumsiness
    author's experience of, 5, 8, 17,
        145–47
    kinesthetic perception deficit and,
        145–47, 149, 153, 221
    symbolic thinking deficit and, 97
Cognitive classroom, 161
Cognitive control, 64
Cognitive exercises, 213, 216
    author's personal use of, 7–8, 41–44,
        147–48, 162, 188
    clock-reading (*see* Clock-reading
        exercise)
    development of, 33–35, 162
    failure to recognize value of, 9
    permanent gain from, 34
    principles of, 33–34
    proficient performance of, 33
    for schizophrenia, 14
Cognitive function, 33, 43
Cognitive Goldilocks, 92
Cognitive Neuroimaging Laboratory, 92
Cohen, Jonathan, 64
Cohen, Joshua, 161, 162, 187–89
    death of, 187
    marriage to author, 39–40
Columbia University, 12

Compensations, xv, 9–10, 210–11
    artifactual thinking and symbolic
        thinking, role in, 96–97
    for auditory speech discrimination
        deficit, 203
    for eye tracking when reading, 123
    for kinesthetic perception deficit, 147,
        148, 154–55
    for memory-for-information deficit,
        180
    for object recognition deficit, 107–108
    for predicative speech deficit, 219
    for quantification deficit, 191, 195–96,
        197
    for reading dysfunction (technology
        for), 125–26
    for spatial reasoning deficit, 166
Con artists, vulnerability to, 25, 46
*Conditioned Reflexes and Neuron
    Organization* (Konorski), 13
Cortical region of the brain, xv, 104, 109,
    110, 112, 168
Cursive writing, 132, 133
Curtis, Elinor, 153–55
Cynically Yours (Peeters), 125

Dalton, Nicolas, 77, 78, 84–86, 87
Dalton, Teresa, 84–86
Davies, Stuart, 67–69, 139–40
Dawes alphabet, 136
Day, Tanya, 77, 78, 79–83, 84, 85, 87, 198
Decade of the Brain, 12
Dehaene, Stanislas, 120, 193–94
Delaying gratification, 62–63
Demosthenes, 152
Dendrites, 30–31, *31*
Depth perception, 140–41
Distinguishing phonemic signs, 203
Doidge, Norman, xiii–xvii, 10, 13, 14, 29,
    213, 215
Double negatives, 25
Douglass, Frederick, 117
Driving
    artifactual thinking deficit and, 101
    kinesthetic perception deficit and, 147,
        155–56
    spatial reasoning deficit and, 170, 175

Drooling, 153
Dysgraphia, 137–39
Dyslexia, 118. *See also* Reading
    dysfunction
    Italian language and, 118

Eaton, Howard, xvii
"Effects of Environmental Complexity and
    Training on Brain Chemistry and
    Anatomy" (Rosenzweig), 29
Effortful processing, 33
Emotional intelligence, 93, 102
"Enriched" environments, 29, 31, *32*
Épredicativeí aspect of speech, 77
Evans, Joseph, 62
Excitatory signals, 30
Executive functions, 58, 62, 92, 225
Exercise and the brain, 13
Expressive language disorders, 78
External speech, 78, 86–87
Eye patch (for cognitive exercise), 136
Eye tracking problems, 101, 123, 129,
    134, 137, 218

Face blindness, 103–6. *See also* Object
    recognition deficit
Facial expressions, inability to read. *See*
    Artifactual thinking deficit
Fadakis, Helen, 190
Farah, Martha, 108, 193
Fields, W. C., 171
Figurative language, difficulty
    understanding, 45–46. *See also*
    Metaphors, difficulty understanding
Fixler, Michoel, 128
Fixler, Reva, 128
Flash cards, 7–8, 19
Fog
    author in, xiv, 4, 23–27
    dispelling of, 41–44
    Zazetsky in, 2, 6
Food aversions, 49, 52
Foreign languages, difficulties learning,
    127–29, 219
Freeman, Abbie, 94
Freeman, Hannah, 94–95
Freeman, Nathaniel, 93–97

Friends, difficulty making, 26, 49, 84, 93.
    See also Social interaction
Frontal eye field (FEF), *119*
Frontal lobes, 62, *225*
    artifactual thinking deficit and, 96,
        101
    author and Zazetsky, shared strength
        in, 4
    Broca's speech pronunciation and, 219
    reading and, *119*
    right frontal lobe, loss of, 92
    symbolic thinking deficit and, 57, 91
Frontal lobotomy, 62
Functional magnetic resonance imaging
    (fMRI), 118
Furtado, Chris, 50–51
Fusiform face area (FFA), 104, *105*
Fusiform gyrus, 104, *119,* 228

Gage, Fred H., 12
Garland, Devorah, 140, 141
Gemara, 55
Genesi, Adriano, 77, 78, 86–87
Getting lost, 5, 8, 42, 165–66, 173–75
"Getting Lost: A Newly Discovered
        Developmental Brain Disorder,"
    168
Gettysburg Address (writing exemplar),
    133
Glial cells, 30–31
Goodman, Annette, 129–30, 137–38,
    180–83
    memory-for-information deficit of,
        113, 115, 177–78, 180
    object recognition deficit of, 112–15
Goodman, Avital
    memory-for-information deficit of,
        180–83
    reading dysfunction and, 129–30
Graham, Jessica, 99–102
Graham, Laura, 99
Grant, Jason, 208
Graphemes, 120
Gratification, delaying, 62–63
Gray matter of the brain, 30, 100
Greene, Anthony J., 178–79
Gretzky, Wayne, 96, 170

Handwriting, 133, *133,* 134, 135, 139,
    141–43, *142,* 185, 218. *See also*
    Writing
Harvard University, 11, 108
Hebb, Donald, 29
Hebb's rule, 29
Hebrew language, 127–29, 130
Heights, fear of, 147, 156
Higher Cortical Functions in Man (Luria),
    101, 180, 203
Hippocampus, 30, 168–69
Holmes, Dana, 202
Horizontal intraparietal sulcus (hIPS),
    193–94, 228
Horowitz, Alexandra, 177
Howie, Sherri Lane, 52
Hughes, Ted, 208
Human Brain and Psychological Processes
    (Luria), 193
Humor. *See* Jokes and humor

I'll Fight On (Zazetsky), 3
Imrie, Karen, 69–72
Inhibitory signals, 30
In Search of Memory (Kandel), 178
Institute of Cognitive Neuroscience, 194
Intelligence/IQ
    artifactual thinking deficit and, 102
    memory-for-information deficit and,
        183
    predicative speech deficit and, 78, 83
    symbolic thinking deficit and, 70, 102
    working memory compared with, 64
Internal speech, 78, 86–87, 219
*Iron Man, The: A Children's Story in Five
    Nights* (Hughes), 208

Jackson, John, 125–27
James, William, 49
Jennifer (girl with quantification deficit),
    196–97
Jewish Educational Center (JEC), 137,
    138, 196
Jobs/work
    artifactual thinking deficit and, 98
    auditory speech discrimination deficit
        and, 205

kinesthetic perception deficit and, 157
object recognition deficit and, 111
predicative speech deficit and, 83
quantification deficit and, 195
spatial reasoning deficit and, 171
symbolic thinking deficit and, 59–60,
    61, 67–68, 72
symbol relations deficit and, 55–56
Johnson, Jeremy, 117–18, 122–24
Jokes and humor, 25, 42, 45, 52, 73
Jugglers, brains of, 30

Kandel, Eric, 12, 29, 178
Kandl, Josh, 137–39
Kandl, Rose, 33–34, 137–39
Karoly, Aliza, 48–49, 51
Karoly, Zachary
    auditory speech discrimination deficit
        of, 205
    symbol relations deficit of, 47–52
Keenan, Julian, 92
Keller, Helen, xiv
Kempermann, Gerd, 12
Kinesthetic perception deficit, 145–59
    author's, 5, 6, 16, 145–48, 172
    of author's brother, 172–73
    Brodmann area corresponding to, 228
    cognitive exercises for, 153–55, 159
    description of, 220–21
    to experience, 150
    features of, 148–50, 157
    spatial reasoning deficit and, 156–57,
        172–73
Kinesthetic speech disorder, 152–53, 221
    to experience, 152
Kinetic melody, 132–33
Kinetic stutter, 133
Kirk, Samuel, 5
Konorski, Jerzy, 13
Kosslyn, Stephen, 108–9
Kuhn, Thomas, 10–11

Language development, 77. *See also* Speech
Lasorda, Tommy, 96
Lassen, Niels, 193
Laziness, 71, 183, 209
Leaky sieve metaphor, 179, 180

Learning disabilities
  author's definition of, 32
  coining of term, 5
  cost of, 212–13
  impact of, 188–89, 207–14
  incidence of, 211
  movement to end use of term, xvi
  seen as immutable, 9–10
  stigma attached to, 210, 212, 214
Left brain hemisphere
  adaptation to handle written symbols,
    110
  auditory speech discrimination deficit
    and, 120, 201, 219
  language development and, 77
  memory-for-information deficit and,
    218
  motor symbol sequencing deficit and,
    132, 136
  reading dysfunction and, 110, 123
  symbolic thinking and, 57
Left-right distinction, problems with, 5,
    147, 172
Lemieux, Mario, 170
Lesion studies, 193
Levertov, Denise, x
Lexical memory deficit, 220
Lindholm, Kirstin, 159
"Linking Mind, Brain and Education"
  (Ronstadt and Yellin), 11
Lobes of the brain, *225*
Lobotomy, 62
London cab drivers, brains of, 30, 169
Low muscle tone, 158
Luria, Aleksandr Romanovich, xvi, xvii,
    6–7, 32, 33, 35, 39, 47, 57, 58, 71,
    86, 87, 147, 155, 162, 165, 187,
    215
  on afferent motor aphasia, 153
  definition of thinking, 59
  on kinetic melody, 132–33
  on memory problems, 180
  on mental arithmetic problems, 193
  on right frontal lobe lesions, 101
  on right-hemisphere functions, 91
  on semantic aphasia, 46
  on speech deficits, 77–78, 80, 203

  on writing difficulties, 132–33, 134,
    136
  Zazetsky and, 3–4
Lyon, G. Reid, 122

Maguire, Eleanor, 168–69
Malacarne, Michele Vincenzo, 13
"Man of Letters, A" (Sacks), 109
Man Who Mistook His Wife for a Hat, The
  (Sacks), 149
*Man with a Shattered World: The History of
  a Brain Wound* (Luria), 3–4, 6
Maps, 166, 167, 168, 169–70, 173–74,
    175, 176
Marcinkowski, Jill, 124, 125
Math difficulties. *See also* Quantification
    deficit; Supplementary motor
    deficit
  author's, 4, 8, 19–21, 24, 41–42, 43
  motor symbol sequencing deficit and,
    131, 135, 138
  symbolic thinking deficit and, 58, 65
  symbol relations deficit and, 45, 53,
    54, 218
Mawhinney, Janice, 106–9, 113–14
McGill University, 13
Mechanical reasoning deficit, 222
Meditation, 30, 108–9
Melville, Herman, 24
Memory
  author's strengths in, 5, 24, 25
  dispersal of, 178–79
  lexical, 220
  muscle, 132, 133, 135
  predicative speech deficit and, 78
  semantic, 178
  sensory, 150, 151, 153, 154
  working, 64–65
Memory-for-information deficit, 177–85,
    204
  cognitive exercises for, 180, 182
  description of, 218
  to experience, 179
  features of, 177–78
  object recognition deficit and, 113, 115
  Zazetsky's, 2
Merzenich, Michael, 14

Metaphors, difficulty understanding, 23, 35, 45
Method of loci, 169
Michael (author's friend), 41
Middle temporal region of the brain, *119*
Midkiff, Carol, 138, 207
Miller, Earl, 64
Miller Analogies Test, 43
Mind, Brain, and Education (journal), 11
Mind, Brain, and Education Institute, 11
Mischel, Walter, 62
MIT, 64
Moats, Louisa C., 122
Moby Dick (Melville), 24
Montclair State University, 92
"Mother's Observations of Her LD Daughter, A" (Whitten-Day), 79–80
Motor homunculus, 158, *158*
Motor plan, 132, 135–36
Motor symbol sequencing deficit, 131–43
    Brodmann area corresponding to, 228
    cognitive exercises for, 135–37, 140
    depth perception and, 140–41
    description of, 217–18
    to experience, 132
    features of, 132, 135
    kinetic melody, 132–33
    memory-for-information deficit and, 185
    reading and, 120, 122, 123, 129, 130, 132, 134, 218
    stereopsis and, 140–141
Multitasking, 63
Muscle memory, 132, 133, 135

Narrow visual span, 221–22
Neocortex, 62
Neuman, Brocha, 55
Neurogenesis, 12
Neuroimaging studies, 14, 118, 193
Neurological fatalism, 10
Neurons, 30–32, *31, 32*
Neuroplasticity, 9–14, 29–32, *32,* 137, 216
    activity-dependent, 14, 33
    current applications of principle, 13–14

dark ages of, 29
differential stimulation and, 30
Doidge on, xiii–xvii, 10, 13, 14, 29, 213, 215
gain-loss myth, 184
recent acceptance of concept, 12
resistance to concept, 9–12, 14
self-directed, 13
structural underpinnings of, 30–32
Neurotransmitters, 30, 31
"New Nerve Cells for the Adult Brain" (Kempermann and Gage), 12
New Yorker (magazine), 103, 109
Nobel Prize winners, 12, 29, 36
Numbers, difficulties with. *See* Math difficulties; Quantification deficit; Supplementary motor deficit
Number Sense, The (Dehaene), 193

Object recognition deficit, 103–15
    Brodmann area corresponding to, 228
    cognitive exercises for, 112
    description of, 222
    to experience, 109
    features of, 104, 110–11
Obsessive-compulsive disorder (OCD), 13
Occipital complex, right lateral, 104
Occipital face area (OFA), 104, *105*
Occipital lobes, 47, *225*
    functions associated with, 6
    narrow visual span and, 221
    reading and, *119*
Occipital-parietal-temporal juncture, 7
    functions associated with, 6
    symbol relations deficit and, 46–47, 49–50, 218
Occipital-temporal area of the brain
    object recognition deficit and, 110
    reading and, *119,* 120–21
    symbol recognition deficit and, 120–21, 220
Ontario Institute for Studies in Education (OISE), 38–39
O'Toole, Kathleen, 184–85
O'Toole, Maureen, 184–85

Paradigm shifts, 10–11
Paraphasia, semantic, 24
Parietal lobes, 47, *225. See also* Occipital-
     parietal-temporal juncture
   functions associated with, 6
   kinesthetic perception deficit and,
      149, 151
   quantification deficit and, 223
   reading and, *119*
   spatial reasoning deficit and, 168
Parkinson's patients, 133
Pearce, Madison, 131, 136
Peeters, Marcel, 124–25
Peirson, Andrea, 123, 153, 154–55,
     171–72
Penfield, Wilder, 62, 92
Performance anxiety, 210
Phonemes, 118, 120, 123, 127, 153
Phonetic languages, 118
Phonological processing, 120
Piaget, Jean, 38
Pile person, 168, 170–71, 175
Plastic paradox, 10
Poetry, 35
Positional apraxia, 153
Posterior parietal cortex, 169
Posttraumatic stress disorder, 13–14
Predicative speech deficit, 77–87, 217
   cognitive exercises for, 81, 84–85, 87
   description of, 219
   external speech and, 78, 86–87
   features of, 77–78
   internal speech and, 78, 86–87, 219
Prefrontal cortex, *225*
   ADHD and, 55, 100
   artifactual thinking deficit and,
      91–93, 99, 101, 102, 221
   cognitive control and, 64
   critical role in cognition, 58, 62–65
   delaying gratification and, 62–63
   executive functions, 58, 62, 92, 225
   location in brain, 62, *225*
   schizophrenia and, 14
   symbolic thinking deficit and,
      57–59, 62–65, 70, 71, 75, 91,
      220
   working memory and, 64–65

Premotor area of the brain, 135, 136–37,
     141, 217, 228
   functions associated with, 132, 133–34
   Luria on damage-related problems, 133
Primary auditory cortex, *119*
Primary motor area of the brain, *119*
   Brodmann area corresponding to, 228
   description of deficit, 223
   kinesthetic perception deficit and,
      157–59
   location of, *158*
Primary visual cortex, 136
Procrastination, 70
Proprioception, 149
Prosopagnosia. *See* Face blindness

Quantification deficit, 191–200, 201
   Brodmann area corresponding to, 228
   cognitive exercises for, 198–99
   description of, 223
   to experience, 194
   features of, 194–95
Queen Mary Public School, 20

Ramón y Cajal, Santiago, 12
Raven's Progressive Matrices, 83
Rayne, Heather, 52–55
Reading dysfunction, 110, 117–30
   auditory speech discrimination deficit
      and, 203
   author's, 5, 19, 21
   cognitive exercises for, 120, 123, 128
   mechanical aspect of, 129
   motor symbol sequencing deficit and,
      120, 122, 123, 129, 130, 132, 134,
      218
   reading aversion stemming from, 129
   Zazetsky's, 3
Reading in the Brain (Dehaene), 120
Refrigerator dysfunction (object
      recognition deficit), 111
Reversal of letters and numbers, 20, 45,
     218
Right brain hemisphere, 91
   abstract reasoning deficit and, 222–23
   artifactual thinking deficit and, 91
   intuition and, 34–35

mechanical reasoning deficit and, 222
object recognition deficit and, 104, 110, 222
Ronstadt, Katie, 11
Rosenzweig, Mark, 7, 29–30, 31, 33

Sacks, Oliver, 103–5, 109, 149
Salk, Jonas, 34
Sanders, Gabriela, 65–67
Saturday Night (magazine), 215
Schizophrenia, 14
Scholastic Aptitude Test (SAT), 63
Schwartz, Jeffrey, 13
Science Daily (magazine), 168
Scientific American Mind (magazine), 178
Secondary associative area of the brain, *119*
Self-awareness, 92–93
Self-directed neuroplasticity, 13
Semantic aphasia, 46
Semantic memory, 178
Semantic paraphasia, 24
Sensory feedback, 148, 150, 153–54, 156
Sensory homunculus, 151, *151*, 158
Sensory memory, 150, 151, 153, 154
Sewing, 167, 170
Shapiro, Claire
   artifactual thinking deficit of, 97–98
   spatial reasoning deficit of, 173–75
Shaywitz, Bennett A., 118
Shaywitz, Sally E., 118
Shepherd, Glenn, 131, 135, 136
Shoelaces, difficulty tying, 147
60 Minutes (television program), 41
Social interaction. *See also* Friends, difficulty making
   artifactual thinking deficit and, 90–98
   predicative speech deficit and, 82
   symbol relations deficit and, 26
Somatosensory cortex, *105*
   kinesthetic perception deficit and, 149, 151–52, 153–54, 158, 220–21
   reading and, *119*
Song lyrics, difficulties understanding, 204
Sound-symbol correspondence, 120, 130

Spatial reasoning deficit, 165–76
   author's, 146–47, 165–68, 170–71, 175–76
   of author's mother, 42
   Brodmann area corresponding to, 228
   chess and, 170
   cognitive exercises for, 165–66, 174
   description of, 222
   to experience, 169
   features of, 166–68, 170
   gender differences in, 172
   hockey and, 170
   kinesthetic perception deficit and, 156–57, 172–73
   video games and, 172
Speech. *See also* Auditory speech discrimination deficit; Kinesthetic speech disorder; Predicative speech deficit
   author's father's problems with, 42–43
   ëpredicativeí aspect of, 77
   motor symbol sequencing deficit and, 132, 134, 137, 218
   Zazetsky's problems with, 3
Spelling problems, 124–25
   auditory speech discrimination deficit and, 203
   motor symbol sequencing deficit and, 132, 134–35, 218
   symbol recognition deficit and, 110
Sperry, Roger Wolcott, 36
Sports, 146–47, 157, 170
Stanford Achievement Test, 129
Stanford marshmallow experiment, 62–63
Stereopsis, 140–41
Structure of Scientific Revolutions, The (Kuhn), 10–11
Subcortical region of the brain, xv, 55, 62, 168
Suicidal ideation, 6, 26, 208
Superior temporal area of the brain, *119*, 120, 201, 219
Superior temporal cortex, *105*
Superior temporal lobe, 127
Superior temporal sulcus, 104
Supplementary motor deficit, 192–93, 223
   Brodmann area corresponding to, 228

Symbolic thinking deficit, 57–75, 100,
        217
    artifactual thinking deficit and, 91,
        96–97, 102
    attentional difficulties and, 55, 100
    Brodmann area corresponding to, 228
    cognitive exercises for, 58–59, 60–61,
        71, 102
    description of, 220
    executive role of, 58, 59, 62
    to experience, 57
    features of, 58–59, 63
    intelligence/IQ and, 70, 102
Symbol recognition area of the brain, 120,
        122, 128, 130
Symbol recognition deficit, 120, 121,
        122, 123, 128, 130
    Brodmann area corresponding to, 228
    to experience, 121
Symbol relations deficit, 45–56
    auditory processing problem and, 49
    author's, 23–27, 45, 147, 188
    Brodmann area corresponding to, 228
    cognitive exercises for, 50, 51, 52, 55
    con artists, vulnerability to and, 25,
        46
    description of, 218
    to experience, 46
    features of, 45–47
    hallmark of, 47, 188
    rigidity and, 24, 26, 52
Synapses, 29–30, 31, 201
Synaptic plasticity, 29
Syntax, 78

Talmud, 55
Taylor-Wright, Ian, 52, 136
Teitz, Eliyahu, 137, 138, 196
Telephone conversations, discomfort with,
        203–4
Telling time, 2, 4, 6, 7–8, 218. See also
        Clock-reading exercise
Temporal lobes, 47, 225. See also
        Occipital-parietal-temporal
        juncture; Occipital-temporal area
        of the brain
    anterior, 105

functions associated with, 6
    memory problems and, 180, 218
    reading and, 119
Theory of mind, 92–93
Thinking
    Luria's definition of, 59
    predicative speech deficit and, 78
Three-dimensional objects, visualizing,
        167, 172, 173, 175
Through the Looking-Glass (Carroll), 103,
        191
Time, difficulties with. See Quantification
        deficit; Telling time
Toronto Star, 106, 107
Tracing exercise, 136, 153–55, 174
Traumatic Aphasia (Luria), 77
Tulloch, Ann
    auditory speech discrimination deficit
        of, 201, 203–5
    motor symbol sequencing deficit of,
        140–41
    quantification deficit of, 197–200,
        201
Typing, 145–46, 148, 150

UCLA School of Medicine, 13
Uncertainty, sense of, 47, 188
Unitarian Fellowship, 15, 37
University College London, 168–69, 194
University of California at Berkeley, 7
University of Guelph, 37–38
University of Pennsylvania, 193
University of Wisconsin-Milwaukee, 178

Vandermeer, Johanna, 57–61
Vanessa (girl with artifactual thinking
        deficit), 89–90
Video games, 172
Vinogradov, Sophia, 14
Visual Agnosia (Farah), 104
Visual associative area of the brain, 119
Visual word form area of the brain, 121.
        See also Occipital-temporal area of
        the brain
Vocabulary, 80, 82
Vogeley, Kai, 92
Vygotsky, Lev, 86, 87

Wall metaphor, 57
Wechsler Intelligence Tests, 167
Welsh language, 121–22
Wepman's Auditory Discrimination Test, 202
WFMU-FM (radio station), 137
Whitten-Day, Ginny, 79–83
Wide Range Achievement Test, 43
Winters, June, 72–74, 155–57
Winters, Robert, 72–74, 155
Woodcock-Johnson Picture Recognition test, 115
Word blindness, 121
Work. *See* Jobs/work
Working Brain, The (Luria), 77, 91, 132–33
Working memory, 64–65
Writing. *See also* Handwriting; Motor symbol sequencing deficit
    author's progress in, 42
    cursive, 132, 133
    development of, 135
    dysgraphia, 137–39
    kinesthetic perception deficit and, 150, 153, 154
    pain caused by, 139, 143
    predicative speech deficit and, 78

symbolic thinking deficit and, 66, 69
Zazetsky's problems with, 2–3

Yale Center for Dyslexia and Creativity, 118
Yellin, Paul, 11
YMCA Vocational Counselling Centre, 60, 162
"You Look Unfamiliar" (Sacks), 103
You Mean I'm Not Lazy, Stupid or Crazy? (Kelly), 71
Young, Alex (brother), 15, 172–73
Young, Donald (brother), 8, 15, 16, 21, 161, 167
Young, Greg (brother), 15
Young, Jack (father), 15–16, 21, 42–43, 165
Young, Mary (mother), 15, 19–20, 27, 42
Young, Will (brother), 15
Young family Christmas newsletter, 27, 39
Young Israel of Ocean Parkway (synagogue), 163

Zazetsky, Lyova, 189, 215
    author's identification with, 5, 6–7, 26
    symbol relations deficit of, 45
    symptoms of, 1–4

ABOUT THE AUTHOR

Barbara Arrowsmith-Young holds a B.A.Sc. in child studies from the University of Guelph and a master's degree in school psychology from the University of Toronto (Ontario Institute for Studies in Education).

The genesis of the Arrowsmith Program of cognitive exercises lies in Barbara Arrowsmith-Young's journey of discovery and innovation to overcome her own severe learning disabilities. Diagnosed in first grade as having a mental block, which today would be identified as multiple learning disabilities, she read and wrote everything backward, had trouble processing concepts, regularly got lost, and was physically uncoordinated. She eventually learned to read and write but continued to have difficulty in school with specific aspects of learning. Her unique combination of severe learning disabilities and intellectual gifts was the driving force in her development of the suite of cognitive exercises she created to strengthen specific functions of the brain. These now constitute the Arrowsmith Program and are used in schools in Canada and the United States.

As the director of Arrowsmith School and the Arrowsmith Program, she continues to develop programs for students with learning disabilities. Her vision is that all students struggling with learning disabilities will have the opportunity to benefit from cognitive programs based on the principles of neuroplasticity to change the brain's capacity to learn.

penguin.co.uk/vintage